VOLUME ONE HUNDRED AND EIGHTEEN

ADVANCES IN COMPUTERS

Durable Phase-Change Memory Architectures

VOLUME ONE HUNDRED AND EIGHTEEN

ADVANCES IN COMPUTERS

Durable Phase-Change Memory Architectures

Edited by

MARJAN ASADINIA
University of Arkansas, Fayetteville, AR, United States

HAMID SARBAZI-AZAD
Sharif University of Technology and Institute for Research in Fundamental Sciences (IPM), Tehran, Iran

Academic Press is an imprint of Elsevier
50 Hampshire Street, 5th Floor, Cambridge, MA 02139, United States
525 B Street, Suite 1650, San Diego, CA 92101, United States
The Boulevard, Langford Lane, Kidlington, Oxford OX5 1GB, United Kingdom
125 London Wall, London, EC2Y 5AS, United Kingdom

First edition 2020

ISBN: 978-0-12-818754-8
ISSN: 0065-2458

For information on all Academic Press publications
visit our website at https://www.elsevier.com/books-and-journals

Publisher: Zoe Kruze
Acquisition Editor: Zoe Kruze
Editorial Project Manager: Peter Llewellyn
Production Project Manager: James Selvam
Cover Designer: Alan Studholme

Typeset by SPi Global, India

Contents

Preface

Traditionally, *Advances in Computers*, the oldest series to chronicle of the rapid evolution of computing, annually publishers several volumes, each one typically comprised of four to eight chapters, describing new developments in the theory and applications of computing.

The 118th volume is an eclectic volume inspired by recent advances in memory technology in general and more specifically on Phase Change Memory (PCM) technology as potential replacement for Dynamic Random Access Memory (DRAM). The volume is a collection of six chapters as follows:

Chapter 1 is an introductory chapter that is intended to introduce the Phase Change Memory (PCM) technology as a promising technology to replace DRAM. PCM offers fast access, negligible leakage energy, superior scalability, high density, and operating in both Single-Level Cell (SLC) and Multi-Level Cell (MLC) storage levels without imposing large storage overhead. However, these advantages come at the expense of lower write endurance and its low resilience to soft errors.

This chapter gives an overview of non-volatile memory technologies followed by a discussion about memory hierarchy in modern computers. It also introduces the PCM technology maturity.

Chapter 2 is intended to provide the necessary background information about PCM from historical as well as material physics perspective. Characteristics of PCM are enumerated and discussed. Memory cell, array design, and multi-level cell PCM organizations are articulated. Finally, the shortcomings of PCM are elaborated.

Chapter 3 starts with a discussion about the future of the memory system as a hybrid of PCM and DRAM arrays leading to a PCM only memory organization. Wear-leveling technique that spreads writes uniformly over memory space is discussed and its security vulnerability is addressed. The chapter also overviews several recent security aware wear-leveling techniques.

Chapter 4 focuses on cell wear-out and PCM main memory degradation. Redirection or correction schemes as potential solutions are studied and a new technique called on-demand page paired PCM memory system is proposed (OD3P). In a nutshell, OD3P technique increases memory durability by redirecting failed pages to healthy pages. Increased durability comes at the expense of gradual memory capacity degradation.

Chapter 5 concentrates on how to: (i) prolong the lifetime of a PCM device, (ii) reduce the write rate to PCM cells, and (iii) handle cell failures when hard faults occur by application of the so-called byte-level shifting mechanism. In addition, it demonstrates that the MLC capability of PCM and manipulation of the data block to recover faulty cells can also be used for error recovery purposes.

Finally, Chapter 6 argues that resistance drift is one of the challenging issues for MLC PCM and in response to this challenge proposes the so-called Variable Resistance Spectrum MLC PCM that improves energy consumption, latency and reliability while maintaining the capacity advantage of PCM.

I hope that readers find this volume of interest, and useful for teaching, research, and other professional activities. I welcome feedback on the volume, as well as suggestions for topics of future volumes.

Ali R. Hurson
Missouri University of Science and Technology
Rolla, MO, United States

CHAPTER ONE

Introduction to non-volatile memory technologies

Marjan Asadinia[a], Hamid Sarbazi-Azad[b]
[a]University of Arkansas, Fayetteville, AR, United States
[b]Sharif University of Technology and Institute for Research in Fundamental Sciences (IPM), Tehran, Iran

Contents

Abstract

Dynamic Random Access Memory (DRAM) has been the leading main memory technology during the last four decades. In deep submicron regime, however, scaling DRAM comes with several challenges caused by charge leakage and imprecise charge placement. Phase Change Memory (PCM) technology is known as one of the most promising technologies to replace DRAM. Compared to competitive non-volatile memories like NAND Flash, Spin Transfer Torque random-access memory (STT-RAM), Magnetoresistive random-access memory (MRAM), PCM benefits from best attributes of fast random access, negligible leakage energy, superior scalability, high density, and operating in both Single-level Cell (SLC) and Multilevel Cell (MLC) storage levels without imposing large storage overhead. Unfortunately, density advantage of MLC PCM devices comes at the cost of lower write endurance that results in fast wear-out of memory cells. Adding to it, other preliminary concerns for PCM applicability are related to low resilience to soft errors because of resistance drift, higher latency and energy consumption. To alleviate this issue, recent studies have proposed redirection or correction schemes, but all suffer from poor throughput and latency. None of the techniques proposed in the literature to improve the lifetime and reliability of PCM memories consider the impressive characteristic of PCMs to easily shape-shifting from SLC to MLC storage level using some negligible overhead of read/write circuits. We exploit this remarkable ability to propose some new techniques to improve the durability and reliability of PCMs in this book. This chapter gives an overview of non-volatile memory technologies, memory hierarchy in modern computers, PCM technology maturity and finally introduce the structure of the book and its chapters.

Advances in Computers, Volume 118
ISSN 0065-2458
https://doi.org/10.1016/bs.adcom.2019.09.001

The requirement of memory capacity is huge for most applications including those applications in server space that have a larger working set. Other applications like I/O intensive applications need a high performance storage system. The existing memory requirement as well as the demand for larger memory capacity would grow in the future [1–7]. So, memory hierarchy has an impressive role in determining the overall performance of a computing system and modern computers. In this line, the minimization of the speed gap between the storage system, memory and processor is a pivotal factor. Moreover, since most system's power budget is limited, power consumption of each memory system should be given a high importance [4,8,9]. If there is low power consumption on the memory side, it opens up a wide scope for inclusion of powerful processors.

1. Memory hierarchy and non-volatile memory

Fig. 1 demonstrates the memory hierarchy used in modern computers. It is designed in such a way that layers like "caches" which are closer to the "processor cores" have low latency access compared to those on a farther side which have a longer duration of access time along with higher capacity [3,8,9].

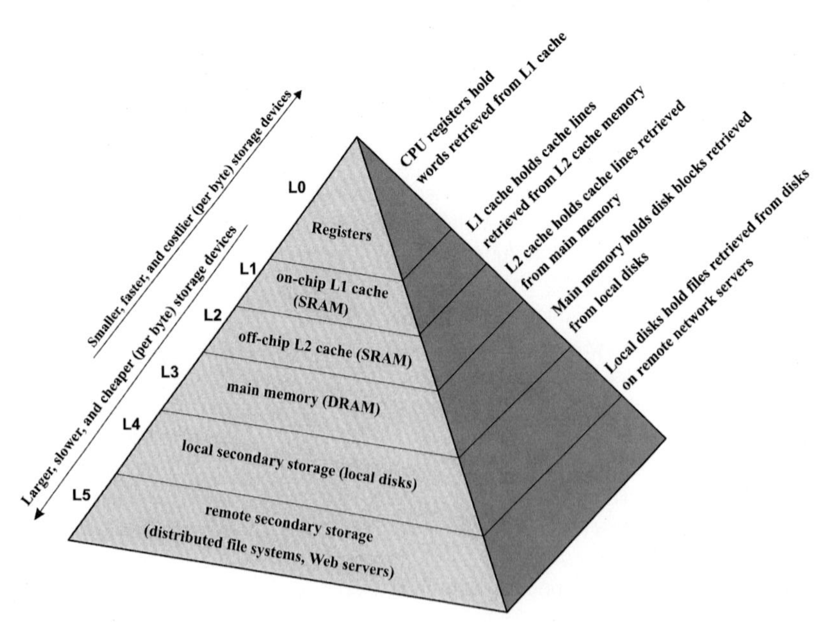

Fig. 1 Memory hierarchy.

There exists a property in the lower layer of this hierarchy known as "Non-volatility" or "Non-volatile Medium." Such a property not only allows for data to be written and read but also includes the ability to hold such retained data for a longer period of time without additional power requirement.

Over many years' caches, main memory and storage have been designed by exploiting basis technologies like Static RAM (SRAM), Dynamic RAM (DRAM) and rotating disks. There have emerged certain challenges from continued use of these technologies:

1. High power consumption and scalability problems in DRAM though it offers high density [5,8].
2. Leakage power which leads to excessive power consumption and low density problems in SRAM while it provides low latency [8].
3. High access latencies which are many times higher than main memory in hard disk though they provide Gigabyte storage at low cost [8].
4. High power consumption in hard disks

The task of flash memory replacing rotating disks in the form of SSD (or Solid State Drives) has provided a huge potential for large boost in performance for storage. However, a wide gap exists in performance between storage and main memory needing to be improved. One way of solving the above challenges is through using NVM or Non-volatile Memory Technology [9–15].

2. Emerging NVM technologies

There are several promising NVM technologies actively being explored in academia and industry. Examples include: NAND Flash, Spin Transfer Torque random-access memory (STT-RAM), Magnetoresistive random-access memory (MRAM), Phase Change Memory (PCM), and others [16–30]. However, the one closest to large-scale production is *PCM* [16–25,31–55]. Fig. 2 shows the current technology trend and promising Non-volatile technology trend.

A PCM cell uses chalcogenide material composed of Ge, Sb, and Te [31–42]. The material has two different states (low resistive crystalline state and high resistive amorphous state). Although PCM has negligible leakage, better scalability, and comparable read speed to DRAM, it suffers from a short lifetime problem [43–60]. A PCM cell can wear-out after a limited number of writes (10^7–10^8 per cell in 32 nm prototypes). It means after a cell reaches its lifetime limit, the cell faces hard error and it is stuck at "1" or "0" state.

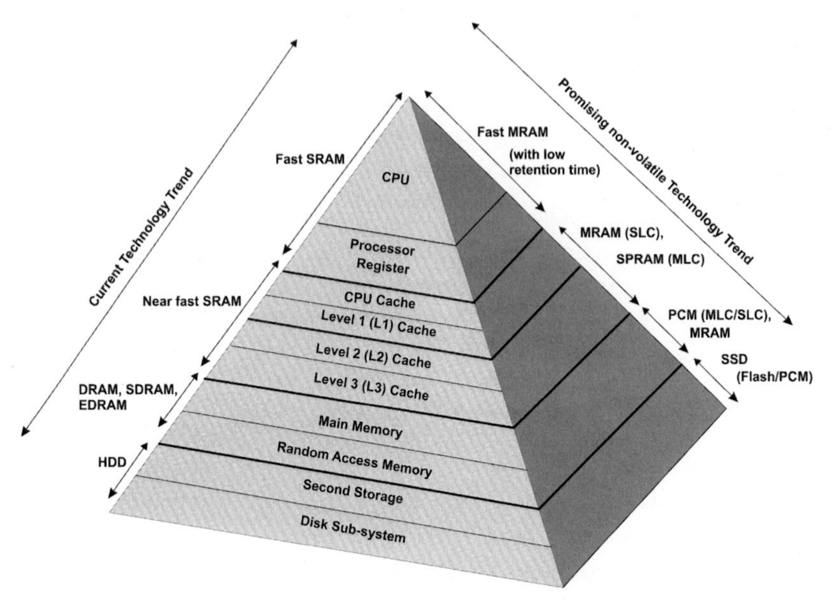

Fig. 2 Memory hierarchy for current technology trend (left), and non-volatile technology trend (right).

Therefore, its limited write endurance and high write latency need to be addressed before this memory technology can be leveraged to build high-performance memory systems [61–99].

Additionally, Phase Change Memory comes with the capacity to store multiple bits of data per memory cell with its latency being closer to that of DRAM. Although MLC-PCM has larger capacity, it suffers from higher latency and higher energy consumption along with soft errors due to having lower resilience to resistance drift [100–110].

Therefore, the solution lies in developing memory structures that have improved latency, energy and reliability of Multi Level Cell-Phase Change Memory without sacrificing capacity.

3. PCM technology maturity

PCM has emerged as a leading contender to take the role of the next generation memory technology. Impressive progress has been made in basic materials, device engineering and chip level demonstrations including the potential for PCM programming and 3D-stacking [17–25]. In addition, there is a large body of knowledge on the failure mechanisms of this

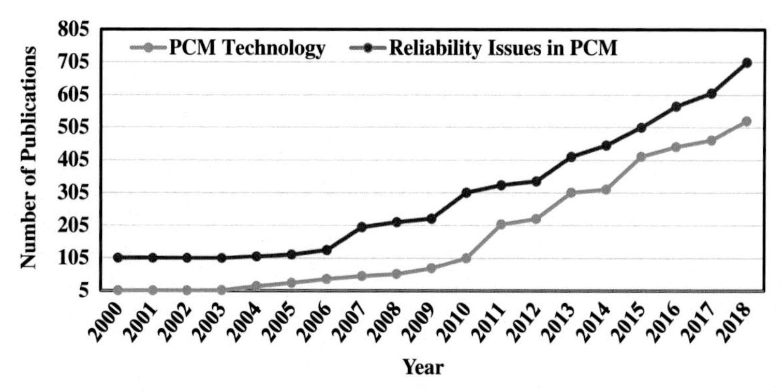

Fig. 3 The number of published articles on PCM Technology taken from IEEE Xplore database for all IEEE, AIP, IET and IBM articles and journals.

technology with active research efforts aimed at suppressing or managing these issues. Fig. 3 shows the number of published articles in IEEE Xplore, ACM and Google Scholar databases concerning the PCM technology, its applicability and reliability during last decade.

4. Contributions

Regarding the performance and write endurance limitations of PCM along with overcoming the reliability shortcomings of MLC-PCM, the current work presents several contributions, briefly described in the following. The goal of all proposed solutions is to improve (or maintain) the performance of a PCM main memory system while addressing the problems of limited endurance, hard errors and drift-induced soft errors.

We first introduce and evaluate the On-Demand Page Paired PCM (OD3P) architecture to address the problem of sudden capacity degradation in PCM main memory systems in presence of severe wear-out of PCM cells. The OD3P architecture relies on this fact that when a failure is detected in a PCM line of cells, there usually exist memory lines that are far from their lifetime limit. We propose to pair PCM lines reaching their lifetime limit with PCM lines receiving low write traffic (cold) and are idle (silent), enabling the former to make use of lifetime of the latter. During recovery, the target PCM line is converted to Multi-Level Cell (MLC) mode to keep the data of both lines. Then, when a line is shifted over a cold-silent line, instead of reducing memory capacity, it worsens read/write latency of the two logical lines (now paired in a physically MLC programmed line).

To mitigate the access latency disadvantage, we further introduce MLC PCM page interleaving with MLC bits to store logical lines with less impact on their access latency.

It is clear the efficiency of line pairing depends on the remaining lifetime and write traffic of the target line. Choosing the most silent target line throughout memory is an important procedure, so we then present some solutions to opportunistically deactivate memory lines to improve overall system performance and lifetime.

To further prolong the lifetime of a PCM device and handle cell failures when hard faults occur, we consider the existing nonuniformity of cells activity within a block and propose BLESS which is a byte-level shifting scheme in order to reduce this nonuniformity. Our method enhances the memory lifetime at two phases: before and after occurrence of hard errors. Before occurrence of hard errors, by uniforming the bit flips in each block, it removes the stress from the hot locations and consequently relaxes the worst-case bit flips rate in the line. After occurrence of hard errors, BLESS changes the storage type of faulty blocks from SLC to MLC. Then, it determines the byte-level fault-mask of the line and shifts the fault-mask in order to find the state that the fault-mask and the shifted form of the fault- mask do not have "1"s in same positions. If such a configuration is found, BLESS shifts the block content accordingly and then combines the shifted block bytes to the original block. The twin blocks are sent as a MLC data block to the write circuit for an MLC write. In fact, our proposed method using a simple and low overhead shifting mechanism not only uniforms the bit flips, but also provides a block-level error recovery technique.

Thereafter, we observe that a line has many healthy cells even though it is marked as faulty. Focusing on these healthy cells and utilizing them can provide lower capacity degradation and longer memory endurance. So the solution we propose is ILP or Intra-line Level Pairing. In ILP, stuck-at faults are recovered by relying on pairing faulty parts of a line with other parts of that line, which still have many healthy cells. By way of pairing, the line can still exist and continue its normal operations until it reaches a stage where the faults can no longer be tolerated. This is enhanced by the idea of using Multi Level Cell (MLC) as a backup storage to store both faulty and target healthy parts when the latter is selected as the target for faulty line.

By doing this we can ensure that faulty part of the faulty line can utilize the lifetime of healthy parts which will then result in improving overall memory lifetime with compromising on performance and energy at SLC level.

All of the prior solutions focused on using MLC PCM and benefits from the density advantage of multilevel cell devices; however, resistance drift is the main burden toward the wide adaption of such devices. To address this principal challenge, we present variable resistance spectrum MLC PCM (VR-PCM) in chapter "Addressing issues with MLC phase-change memory" by Asadinia and Sarbazi-Azad.

It can maintain high capacity and also efficiently handle any drift aware MLC PCM access operations. It operates by relying on the observation that data patterns with various granularities are distributed nonuniformly across memory transactions when running various workloads. By using this technique, VR-PCM modifies PCM resistance spectrum which partition into nonuniform regions. These are then assigned to various binary data patterns as per their occurrence frequency.

5. Organization of the book

After the brief introduction of chapter "Introduction to non-volatile memory technologies", we provide necessary background information on PCM in chapter "The emerging phase-change memory". Then, in chapter "Phase-change memory architectures" we review related work on PCM reliability along with a review on the efficiency and overheads of different schemes. In chapter "Inter-line level schemes for handling hard errors in PCMs", we introduce our first technique: On-Demand Page Paired PCM (OD3P) which is a main memory design based on MLC PCM maintaining high capacity even in the presence of rapid cell wear-outs. Later in this chapter we present Byte-level shifting scheme (BLESS) and try to tolerate the occurrence of hard errors. These mechanisms and various design considerations are discussed in greater detail in chapter "Handling hard errors in PCMs by using intra-line level schemes". As well as this scheme, we also propose two Intra-line Level Pairing methods to recover from stuck-at faults. Presented Intra-line level schemes use statically partitioning a data-block into some partitions and provide strong error correction scheme with minimal storage overhead. Chapter 6 includes an explanation of Variable Resistance Spectrum MLC PCM (VR-PCM) as well as presenting details of this architecture alongside its access circuit providing the required reconfigurability.

References

[1] Lefurgy C, Rajamani K, Rawson F, Felter W, Kistler M, Keller TW: Energy management for commercial servers, *IEEE Comp* 36(12):39–48, 2003.
[2] Lim K, Ranganathan P, Chang J, Patel C, Mudge T, Reinhardt S: Understanding and designing new server architectures for emerging warehouse-computing environments. In *Proc. the 35th Annual Int. Symp. Computer Architecture*; 2008, pp 315–326.
[3] Zhao J, Li S, Yoon DH, Xie Y, Jouppi N, Kiln P: Closing the performance gap between systems with and without persistence support. In *Proc. the 46th Annual IEEE/ACM Int. Symp. Microarchitecture*; 2013, pp 421–432.
[4] Dhiman G, Ayoub R, Rosing T: PDRAM: a hybrid PRAM and DRAM main memory system. In *ACM/EDAC/IEEE Design Automation Conference (DAC)*; 2009, pp 664–669.
[5] Mittal S: A survey of architectural techniques for DRAM power management, *Int J High Perform Syst Arch* 4(2):110–119, 2012.
[6] Udipi AN, Muralimanohar N, Chatterjee N, Balasubramonian R, Davis A, Jouppi NP: Rethinking DRAM design and organization for energy-constrained multi-cores, *ACM SIGARCH Comput Archit News* 38(3):175–186, 2010.
[7] Lim K, et al: Disaggregated memory for expansion and sharing in blade servers, *ACM SIGARCH Comput Archit News* 37(3):267–278, 2009.
[8] Arden W, et al: Semiconductor Industries Association, *Int Technol Roadmap Semiconductors,* 2009. http://www.itrs.net.
[9] Meena JS, Sze SM, Chand U, Tseng T: Overview of emerging non-volatile memory technologies, *Nanoscale Res Lett*: 1–33, 2014.
[10] Li D, Vetter JS, Marin G, McCurdy C, Cira C, Liu Z, Yu W: Identifying opportunities for byte-addressable non-volatile memory in extreme-scale scientific applications. In *Proceedings of IPDPS*; 2012, pp 945–956.
[11] Senni S, Torres L, Sassatelli G, Gamatie A, Mussard B: Emerging non-volatile memory technologies exploration flow for processor architecture. In *Proceedings of ISVLSI*; 2015, pp 460–465.
[12] Mutlu O: Main memory scaling: challenges and solution directions. In *More Than Moore Technologies for Next Generation Computer Design*, 2015, Springer, pp 127–153.
[13] Neale RG, Nelson DL, Moore GE: Non-volatile, re-programmable, read mostly memory is here, *IEEE Trans Electronics* 70:56, 1970.
[14] Shanks R, Davis C: A 1024-bit nonvolatile 15 ns bipolar read-write memory, *Proc IEEE Int'l Conf Solid-State Circuits* 3(7):112–123, 1978.
[15] Lai S, Lowrey T: OUM—a 180 nm nonvolatile memory cell element technology for stand alone and embedded applications, *Elect Devices Meet* 36–54, 2001.
[16] Lee BC, Zhou P, Yang J, Zhang Y, Zhao B, Ipek E, Mutlu O, Burger D: Phase-change technology and the future of main memory, *IEEE Micro* 30(1):131–141, 2010.
[17] Kau DC, et al: A stackable cross point Phase Change Memory. In *Proceedings of IEEE International Electron Devices Meeting (IEDM)*, 2009, pp 27.1.1–27.1.4.
[18] Burr GW, et al: Phase change memory technology, *J Sci Technol* 2(10):1–28, 2010.
[19] Wong HP, et al: Phase change memory, *Proc IEEE Trans Comput* 98(12):2201–2227, 2010.
[20] Qureshi MK, Gurumurthi S, Rajendran B: *Phase Change Memory: From Devices to Systems, Synthesis lectures on computer architecture* 2011, Morgan & Claypool.
[21] Lee BC, et al: Architecting phase change memory as a scalable DRAM alternative. In *Proc. Int'l Symp. Computer Architecture (ISCA '09)*; 2009, pp 2–13. June.
[22] Sarkar J, Gleixner B: Evolution of phase change memory characteristics with operating cycles: electrical characterization and physical modeling, *Appl Phys Lett:* 282–294, 2007.

[23] Im DH, et al: A unified 7.5 nm dash-type confined cell for high performance PRAM device, *Electron Devices Meet*: 1–4, 2008.
[24] Breitwisch M: Novel lithography-independent pore phase change memory, *IEEE Symp VLSI Technol*: 100–104, 2007.
[25] Rajendran B, et al: Dynamic resistance—a metric for variability characterization of phase-change memory, *Electron Device Lett* 30(2):126–129, 2009.
[26] Kgil T, Roberts D, Mudge T: Improving NAND flash based disk caches. In *International Symposium on Computer Architecture (ISCA 08)*; 2008, pp 327–338.
[27] Ban A, Hasharon R: *Wear Leveling of Static Areas in Flash Memory,* U.S. Patent Number 6, 44–49.
[28] Ben-Aroya A, Toledo S: Competitive analysis of flash-memory algorithms. In *Proc. Annual European Symposium*; 2006, pp 100–111.
[29] Gal E, Toledo S: Algorithms and data structures for flash memories, *Proc ACM Comput Surv* 37(2):138–163, 2005.
[30] Xie Y, et al: A novel architecture of the 3D stacked MRAM L2 cache for CMPs. In *Proc. IEEE Symp. High Performance Computer Architecture (HPCA' 09)*; 2009, pp 36–48.
[31] Pohm A, Sie C, Uttecht R, Kao V, Agrawal O: Chalcogenide glass bistable resistivity (ovonic) memories, *Proc IEEE Trans Magn* 6(3):1–17, 1970.
[32] Rao F, et al: Multilevel data storage characteristics of phase change memory cell with double layer chalcogenide films (Ge2Sb2Te5 and Sb2Te3), *J Appl Phys* 46(2):25–37, 2007.
[33] Oh GH, et al: Parallel multi-confined (PMC) cell technology for high density MLC PRAM. In *Proc. IEEE Symp. VLSI Technology*; 2009, pp 220–232.
[34] Shih YH, et al: Mechanisms of retention loss in Ge2Sb2Te5-based Phase-Change Memory. In *Electron Devices Meeting, 2008. IEDM 2008. Proc. IEEE International Electron Devices Meeting*; 2008, pp 1–4.
[35] Zhou J, et al: Formation of large voids in the amorphous phase-change memory Ge2Sb2Te5 alloy, *Phys Rev Lett*: 24–102, 2009.
[36] Friedrich I, et al: Structural transformations of Ge2Sb2Te5 films studied by electrical resistance measurements, *J Appl Phys* 87(9):4130–4134, 2000.
[37] Maimon JD, Hunt KK, Burcin L, Rodgers J: Chalcogenide memory arrays: characterization and radiation effects, *IEEE Trans Nucl Sci* 50(6):1878–1884, 2003.
[38] Jang MH, et al: Structural stability and phase-change characteristics of $Ge_2Sb_2Te_5/SiO_2$ nano-multilayered films, *Electrochem Solid-State Lett* 12(4):34–38, 2009.
[39] Raoux S, et al: Direct observation of amorphous to crystalline phase transitions in nano particle arrays of phase change materials, *J Appl Phys* 102(9):18–23, 2007.
[40] Zhou G, Herman J, Borg JC, Rijpers N, Lankhorst M: Crystallization behavior of phase change materials: comparison between nucleation-and growth-dominated crystallization, *Proc IEEE Opt Data Storage*: 74–76, 2000.
[41] Neale RG, Aseltine JA: The application of amorphous materials to computer memories, *IEEE Trans Electron Devices* 20(2):195–205, February 1973.
[42] Gopalakrishnan K, et al: Highly-scalable novel access device based on Mixed Ionic Electronic conduction (MIEC) materials for high density phase change memory (PCM) arrays. In *Proc. Int'l Symp. VLSI Technology*; 2010, pp 205–217.
[43] Zhong M, Song ZT, Liu B, Wang LY, Feng SL: Switching reliability improvement of phase change memory with nanoscale damascene structure by $Ge_2Sb_2Te_5$ CMP process, *Electron Lett* 44(4):322–323, 2008.
[44] Rajendran B, et al: On the dynamic resistance and reliability of phase change memory. In *Proc. IEEE Symp. VLSI*; 2008, pp 96–107.

[45] Goux L, et al: Degradation of the reset switching during endurance testing of a phase-change line cell, *IEEE Trans Electron Devices* 56(2):354–358, 2009.
[46] Ohta T: Phase change optical memory promotes the DVD optical disk, *J Optoelectron Adv Mater* 3(3):609–626, 2001.
[47] Servalli G, et al: A 45 nm generation phase change memory technology. In *Proc. IEEE Int'l Conf. Electron Devices Meeting (IEDM' 09)*; 2009, pp 1–4.
[48] Kang DH, et al: Two-bit cell operation in diode-switch phase change memory cells with 90 nm technology. In *IEEE Int'l Symp. VLSI Technology*; 2008, pp 98–115.
[49] Hwang YN, et al: MLC PRAM with SLC write-speed and robust read scheme. In *IEEE Int'l Symp. VLSI Technology (VLSIT)*; 2010, pp 201–208.
[50] Lin J-T, Liao Y-B, Chiang M-H, Hsu W-C: Operation of multilevel phase change memory using various programming techniques. In *IEEE Int'l Conf. IC Design and Technology*; 2009, pp 199–202. May.
[51] Nakayama K, Takata M, Kasai T, Kitagawa A, Akita J: Pulse number control of electrical resistance for multi-level storage based on phase change, *J Phys* 40(17):55–67, 2007.
[52] Nirschl T, et al: Write strategies for 2 and 4-bit multi-level phase-change memory. In *Proc. IEEE Int'l Electron Devices Meeting*; 2007, pp 461–464.
[53] Bedeschi F, et al: A bipolar-selected phase change memory featuring multi-level cell storage, *IEEE J Solid-State Circuits* 44(1):217–227, 2009.
[54] Grupp LM, Davis JD, Swanson S: The Harey Tortoise: Managing heterogeneous write performance in SSDs. In *ATC*; 2013. Apr.
[55] Cabrini A, et al: Voltage-driven multilevel programming in phase change memories. In *IEEE MTDT*; 2009.
[56] Chimenton A, Zambelli C, Olivo P, Pirovano A: Set of electrical characteristic parameters suitable for reliability analysis of multimegabit phase change memory arrays. In *Proc. Int'l Non-Volatile Semiconductor Memory Workshop on Memory Technology and Design*; 2008, pp 49–51.
[57] Kim K, Ahn SJ: Reliability investigations for manufacturable high density PRAM. In *Proc. IEEE Int'l Symp. Reliability*; 2005, pp 157–162.
[58] Pirovano A, et al: Reliability study of phase-change nonvolatile memories, *IEEE Trans Device Mater Reliability* 4(3):422–427, 2004.
[59] Lacaita AL, Ielmini D: Reliability issues and scaling projections for phase change non volatile memories. In *Proceedings of IEEE Electronic Devices Meeting (IEDM)*; 2007, pp 157–160.
[60] Close GF, et al: Device, circuit and system-level analysis of noise in multi-bit phase-change memory. In *Electron Devices Meeting (IEDM)*; 2010, pp 51–54.
[61] Asadinia M, Jalili M, Sarbazi-Azad H: BLESS: a simple and efficient scheme for prolonging PCM lifetime. In *Proc. Design Automation Conference (DAC)*, pp 1–6.
[62] Lastras-Montano LA, et al: An area and latency assessment for coding for memories with stuck cells. In *Pro. GLOBECOM Workshops (GCWkshps)*; 2010, pp 1851–1855.
[63] Zhou P, et al: A durable and energy efficient main memory using phase change memory technology. In *Proc. IEEE Symp. High Performance Computer Architecture (HPCA'09)*; 2009, pp 14–23.
[64] Qureshi MK, Srinivasan V, Rivers JA: Scalable high performance main memory system using phase-change memory technology. In *ISCA*; 2009, pp 24–33. June.
[65] Qureshi MK, Franceschini MM, Lastras-Monta LA: Improving read performance of phase change memories via write cancellation and write pausing. In *Proc. IEEE Symp. High Performance Computer Architecture (HPCA'10)*; 2010, pp 1–11. January.
[66] Qureshi MK, et al: Enhancing lifetime and security of PCM based main memory with start-gap wear leveling. In *Proc. IEEE/ACM Int. Symp. Microarchitecture (MICRO'09)*; 2009, pp 14–23. December.

[67] Jalili M, Sarbazi-Azad H: Endurance-aware security enhancement in non-volatile memories using compression and selective encryption, *IEEE Trans Comp* 66(7):1132–1144, 2017.
[68] Hoseinzadeh M, Arjomand M, Sarbazi-Azad H: Reducing access latency of MLC PCMs through line striping. In *International Symposium on Computer Architecture (ISCA)*; 2014, pp 277–288.
[69] Seznec A: *Towards phase change memory as a secure main memory,* Technical report, INRIA, November.
[70] Seong NH, Woo DH, Lee H-HS: Security refresh: prevent malicious wear-out and increase durability for phase-change memory with dynamically randomized address mapping. In *ISCA*; 2010. June.
[71] Seznec A: A phase change memory as a secure main memory, *Proc IEEE Comput Archit Lett*: 57–62, 2010.
[72] Ipek E, Condit J, Nightingale EB, Burger D, Moscibroda T: Dynamically replicated memory: building reliable systems from nanoscale resistive memories. In *ASPLOS*; 2010, pp 3–14.
[73] Asadinia M, Arjomand M, Sarbazi-Azad H: Prolonging lifetime of PCM-based main memories through on-demand page pairing. In *ACM Trans. Design Automation Electronic Systems (ACM TODAES)*; 2015, pp 1–24.
[74] Schechter SE, et al: Use ECP, not ECC, for hard failures in resistive memories. In *Proc. Int. Symp. Computer Architecture (ISCA)*; 2010, pp 141–152.
[75] Shedletsky JJ: Error correction by alternate-data retry, *IEEE Trans Comput* 27(3):678–680, 1978.
[76] Asadinia M, Arjomand M, Sarbazi-Azad H: OD3P: On-demand page paired PCM. In *Proc. Design Automation Conference (DAC)*, , pp 1–6.
[77] Seong NH, et al: SAFER: stuck-at-fault error recovery for memories. In *MICRO*; 2010, pp 115–124.
[78] Yoon DH, et al: FREE-p: protecting non-volatile memory against both hard and soft errors. In *HPCA*; 2011, pp 466–477.
[79] Asadinia M, Sarbazi-Azad H: Using intra-line level pairing for graceful degradation support in PCMs. In *IEEE Computer Society Annual Symposium on VLSI (ISVLSI)*; 2015, pp 527–532.
[80] Wu Q, et al: Using multi-level phase change memory to build data storage: a time-aware system design perspective, *IEEE Trans. Comput.* 62(10), 2013.
[81] Asadinia M, Jalili M, Sarbazi-Azad H: Data block partitioning for recovering stuck-at faults in PCMs. In *International Conference on Networking, Architecture, and Storage (NAS)*; 2017, pp 1–8.
[82] Qureshi MK, et al: Morphable memory system: a robust architecture for exploiting multi-level phase change memories. In *Proc. Int. Symp. Computer Architecture (ISCA)*; 2010, pp 153–162. June.
[83] Jiang L, et al: Improving write operations in MLC phase change memory. In *Proc. IEEE Symp. High Performance Computer Architecture (HPCA)*; 2012, pp 201–210. February.
[84] Azevedo R, Davis JD, Strauss K, Gopalan P, Manasse M, Yekhanin S: Zombie memory: extending memory lifetime by reviving dead blocks. In *Proc. Int. Symp. Computer Architecture (ISCA)*; 2013, pp 452–463. June.
[85] Jiang L, Du Y, Zhang Y, Childers BR, Yang J: LLS: cooperative integration of wear-leveling and salvaging for PCM main memory. In *Proc. Int. Conf. Dep. Sys. and Net. (DSN)*; 2011, pp 221–232.
[86] Jalili M, Sarbazi-Azad H: Captopril: reducing the pressure of bit flips on hot locations in non-volatile main memories. In *Design Automation Test in Europe Conference Exhibition (DATE)*; 2016, pp 1116–1119.

[87] Jalili M, Sarbazi-Azad H: Tolerating more hard errors in MLC PCMs using compression. In *International Conference on Computer Design (ICCD)*; 2016, pp 117–123.
[88] Qureshi MK: Pay-as-you-go: low-overhead hard-error correction for phase change memories. In *Proc. IEEE/ACM Int. Symp. Microarchitecture (MICRO);* pp 318–328.
[89] Dong X, Xie Y: AdaMS: adaptiveMLC/SLC phase-change memory design for file storage. In *Proc. Asia and South Pacific Design Automation Conference (ASP-DAC)*; 2011, pp 31–36.
[90] Hoseinzadeh M, Arjomand M, Sarbazi-Azad H: SPCM: the striped phase change memory. In *ACM Trans. Archit. Code Optim. (TACO)*; 2016, pp 1–25. vo. 12, no. 4.
[91] Zhao M, Jiang L, Zhang Y, Xue CJ: SLC-enabled wear leveling for MLC PCM considering process variation. In *Proc. Annual Design Automation Conference (DAC)*; 2014, pp 1–6.
[92] Zhang Y, Yang J, Memaripour A, Swanson S: Mojim: a reliable and highly-available non-volatile memory system. In *Proc. Int. Conf. Architectural Support for Programming Languages and Operating Systems (ASPLOS)*; 2015, pp 3–18.
[93] Nair PJ, Chou C, Rajendran B, Qureshi MK: Reducing read latency of phase change memory via early read and Turbo Read. In *Proc. Int. Symp. High Performance Computer Architecture (HPCA)*; 2015, pp 309–319.
[94] Jalili M, Sarbazi-Azad H: Express read in MLC phase change memories. In *ACM Trans. Design Automation of Electronic Systems (ACM TODAES)*; 2018, pp 1–33. vol. 23, no. 3.
[95] Maddah R, Melhem R, Cho S: RDIS: tolerating many stuck-at faults in resistive memory, *IEEE Trans Comput* 64(3):847–861, 2015.
[96] Rashidi S, Jalili M, Sarbazi-Azad H: Improving MLC PCM performance through relaxed write and read for intermediate resistance levels, *ACM Trans Archit Code Optim (TACO)* 15(1):1–31, 2018.
[97] Seyedzadeh SM, Maddah R, Jones A, Melhem R: PRES: pseudo-random encoding scheme to increase the bit flip reduction in the memory. In *Proc. Design Automation Conference (DAC'15)*; 2015, pp 1–6.
[98] Wang J, Dong X, Xie Y, Jouppi N: i2wap: improving non-volatile cache lifetime by reducing inter-and intra-set write variations. In *IEEE 19th International Symposium on High Performance Computer Architecture (HPCA'13)*; 2013, pp 234–245.
[99] Fan J, Jiang S, Shu J, et al: Aegis: partitioning data block for efficient recovery of stuck-at-faults in phase change memory. In *Proc. the 46th Annual IEEE/ACM Int. Symp. Microarchitecture*; 2013, pp 433–444.
[100] Asadinia M, Arjomand M, Sarbazi-Azad H: Variable resistance spectrum assignment in phase change memory systems, *IEEE Trans VLSI Systems* 23(11):2657–2670, 2015.
[101] Zhang W, Li T: Helmet: a resistance drift resilient architecture for multi-level cell phase change memory system. In *Proc. IEEE/IFIP Int. Conf. Dependable Systems and Networks (DSN)*; 2011, pp 197–208.
[102] Zhang W, Li T: Characterizing and mitigating the impact of process variations on phase change based memory systems. In *Proc. IEEE/ACM Int. Symp. Microarchitecture*; 2009, pp 2–13. December.
[103] Awasthi M, et al: Efficient scrub mechanisms for error-prone emerging memories. In *Proc. IEEE Symp. High Performance Computer Architecture (HPCA)*; 2012, pp 15–26. February.
[104] Jalili M, Arjomand M, Sarbazi-Azad H: A reliable 3D MLC PCM architecture with resistance drift predictor. In *International Conference on Dependable Systems and Networks (DSN)*; 2014, pp 204–215.

[105] Xu W, Zhang T: Using time-aware memory sensing to address resistance drift issue in multi-level phase change memory. In *Proc. IEEE Symp. Quality Electronic Design (ISQED' 10)*; 2010, pp 356–361.
[106] Joshi M, et al: Mercury: a fast and energy-efficient multi-level cell based phase change memory system. In *HPCA*; 2011.
[107] Ielmini D, Lavizzari S, Sharma D, Lacaita AL: Physical interpretation, modeling and impact on phase change memory (PCM) reliability of resistance drift due to chalcogenide structural relaxation, *Electron Devices Meet*: 23–36, 2007.
[108] Redaelli A, et al: Numerical implementation of low field resistance drift for phase change memory simulations. In *Proc. Non-Volatile Semiconductor Memory Workshop*; 2008, pp 39–42.
[109] Awasthi M, et al: Handling PCM resistance drift with device, circuit, architecture, and system solutions. In *Proc. Non-Volatile Memory Workshop*; 2011, pp 23–35.
[110] Jalili M, Sarbazi-Azad H: A compression-based morphable PCM architecture for improving resistance drift tolerance. In *IEEE International Conference on Application-specific Systems, Architectures and Processors (ASAP)*; 2014, pp 232–239.

CHAPTER TWO

The emerging phase change memory

Marjan Asadinia[a], Hamid Sarbazi-Azad[b]
[a]University of Arkansas, Fayetteville, AR, United States
[b]Sharif University of Technology and Institute for Research in Fundamental Sciences (IPM), Tehran, Iran

Contents

Abstract

This chapter evaluates viewpoints on the Phase Change Memory (PCM) devices and materials entailing multi-level cell (MLC) phase change memory as well as its trade-offs. This chapter lists the main difficulties related to PCMs and possible recommendations to address those challenges. The next chapters introduce some simple techniques to alleviate some of the problems listed here.

1. Introduction

The premise of PCM has its origins in the late 1960s. The concept was introduced in S. Ovshinsky's work on the physics of amorphous and disordered solids [1]. It was suggested by Ovshinsky that there are transitions between the ordered and disordered phases of chalcogenide glasses (alloys include the components from the chalcogen family, such as O, S, Se, Te) that clearly depict forming solid-state memory devices by using different resistivity [1]. It is notable that the transitions between these ordered and disordered phases attain either through optical laser pulses or electrically [2]. This follows demonstration of the first array of PCM in 1970.

Advances in Computers, Volume 118
ISSN 0065-2458
https://doi.org/10.1016/bs.adcom.2019.09.002

A 256 bit "read mostly memory" configured as an array of 16×16 of 'ovonic devices' in succession with integrated silicon diodes [3–5]. PCM uses electronic pulses along with energy-time profile of the applied pulses for programming and determining the memory device's final state. However, a drawback of PCM to become an accepted choice for solid-state memory is that PCM requires higher programming energy compared to DRAM or flash memories.

Fortunately, the energy needed for programming scales is below Pico-Joule range for corresponding active volume of the PCM element. This is achieved once the diameter of active volume goes below 50 nm. [6–9]. The scaling difficulty of flash memory presents an opportunity for greater acceptance of PCM to enter the marketplace and become more widely accepted.

2. PCM materials/device physics

Great variety of materials exists in both disordered amorphous and ordered poly-crystalline or crystalline forms; however, chalcogenide phase change materials (e.g., $Ge_2Sb_2Te_5$, GST in short) are best applicable for memory technological advancement of next generation as they include the below properties:

- *Resistivity difference between amorphous phases and poly-crystalline:* The material has an amorphous form with usual resistivity higher than 10^7 Ω/sq. at room temperature. The resistivity of the material gets reduced during annealing. This is due to changes in the structure of crystal ordering based on the temperature. In fact, a great drop in resistivity is found at about 150 °C (factor of ~100) and 350 °C (factor of ~10) (Fig. 1A). The reason is that the material first crystallizes to a metastable form and then into a stable phase at these two temperatures [10–13].
- *Stable phases at operating temperatures:* The amorphous phase is stable and memory cells can hold their state for longer durations. In some cases, they can hold their state for about 10 years even during higher temperatures of ~85 °C.
- *Fast programming:* The PCM core indicated in Fig. 2 consists of a thin layer of phase change material that is contacted from left and right by metal electrodes. PCM relies on phase changing of the Chalcogenide material (GST) to represent binary information. Timescale of 5–500 ns with electrical pulses is used to achieve transition between both states of the compact memory cell. For RESET operation where a memory

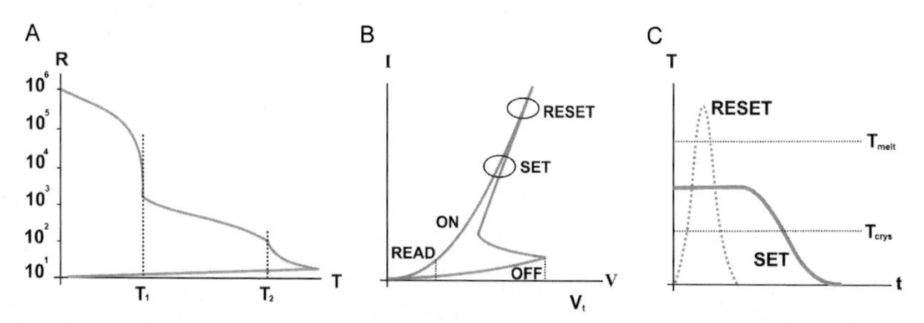

Fig. 1 (A) R-T (normalized) curve of chalcogenide films (it is notable that resistivity of amorphous phase is five to six orders of magnitude higher than the poly-crystalline phase). T_1 and T_2 are the temperatures where the transitions to both phases happen. (B) PCM device behaves like a non-linear resistor and I–V curves showed that in ON state. The device undergoes threshold switching at a critical bias (V_t) in Off state. (C) During SET and RESET programming, ideal thermal profiles generated within the cell.

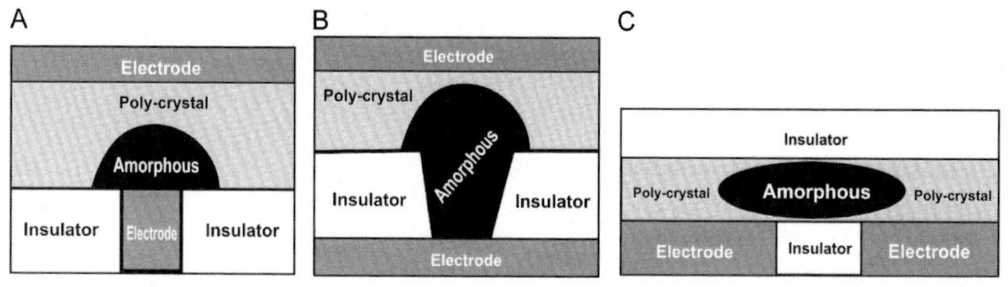

Fig. 2 (A) A thin film of chalcogenide material is contacted by electrodes, with the diameter of the bottom electrode contact (BEC) in the range of 10–50 nm. (B) A nanoscale pore (dimension 10–50 nm) is filled with the chalcogenide material. In the bridge cell (C), a thin layer (5–50 nm thick) of chalcogenide material is deposited on top of patterned electrodes that are few 10s of nm apart.

cell is programmed to high resistance state, the temperature is increased beyond the melting point of 660 °C using electrical pulse in the active volume of chalcogenide material [14–16].

The condition of thermal steady state could be achieved during the same timescale. It is possible to quench the molten volume during a high resistive amorphous state when RESET pulse fall time is just a few nanoseconds.

To program the memory cell to the low-resistance state and SET operation, recrystallize the amorphous material back to the poly-crystalline phase. This requires increasing the temperature to 350 °C. Tight distribution of SET resistances in a cells' collection requires a pulse with a slow ramp down time so cell temperature is reduced gradually. The crystallization speed mainly determines the rate of ramp down of the phase change material, that can either be dominated by growth of crystal front or by nucleation of crystal seeds [17]. The nucleation dominated material entail the commonly used chalcogenide $Ge_2Sb_2Te_5$ [18]. While Te free Ge Sb alloys are the growth dominated materials [19] and can be doped to create combination of the fast crystallization speed with high crystallization temperature, maintaining the expectations for a non-volatile and fast (<100 ns programming) phase change memory technology [20].

- *Large endurance:* Chalcogenide material makes it possible to transit between states of amorphous and poly-crystalline any number of times. Atomic diffusion or segregation may occur at an increased temperature resulting in altering the composition of the alloy. In a worst-case scenario, this creates voids and shorts in a device structure [21,22]. These unimpressive features could be reduced to extend the lifetime of a memory cell to 10^6–10^8 [23] program/erase cycles. Research is then conducted in the device and material engineering resulting in further improvement through changes in programming schemes designs and alloy composition of PCM. This improvement is due to the proof of endurance loss mechanisms, which are depending on time spent by a material in the molten state [24].

A point here to be noted is that the high resistive amorphous phase's reflectivity is poor in comparison to the material's resistive poly-crystalline phase in the 400–800 nm region of the electro-magnetic spectrum [25].

3. Memory cell and array design

Minimizing the current of programming while preserving the resistance of device at reasonable values is one of the main objectives of the phase change element (PCE). This in turn, will help in attaining good write and read speeds. The RESET programming operation requires high current, while SET programming operation is time consuming. The range of 5–20 mA/cm^2 is the programming required in RESET current density. The nanoscale range is used to pattern at least one typical PCE in a physical dimension [26,27]. It is also crucial to reduce the variability related with this characteristic, as it is closely associated with the device resistance and programming current. Techniques of processing are available to obtain nanoscale features regularly and they don't have conventional lithography variability [28–30]. Furthermore, the chalcogenide's thickness and surrounding dielectric films helps in determining the losses of thermal diffusion that further impacts the cooling rate within the active volume and thermal cross-talk between the adjacent cells [31].

The configuration of PCM array is done in a hierarchical manner. Fundamental blocks are built in bit lines and word lines of 1024-4096. There is usually a PCE connected in series with an access device at the connection of each WL and BL. CMOS FET is the most desired choice for an access device [32], because minor changes in the process are essential to alter the features of FETs available in common foundry offerings suitable to the requirements of a memory array. The limitation is the capability of current drive in a FET (of minimum size) is lower than the minimum requirement of PCM programming. The requirement of usage of wider devices results in the large cell area. Therefore, Memory designers exploit BJTs and diodes to be used as access devices due to the fact they offer increased current drive for one unit area in contrast to FETs. BJT or diode based memory arrays come with the advantage of BL/WL contact in each line. However four to eight devices could share it leading to reduce cell size.

The employment of BJTs and diodes has always been demonstrated as access devices for PCM arrays with cell sizes in the range of $5.5F^2$–$11.5F^2$ [33,34]. However, the major disadvantage of this is the additional complexity in the process, mainly because such devices are required to be integrated in crystalline silicon substrates (before the metal levels as well as memory devices themselves are fabricated).

Array configuration and peripheral circuits have a great role in determining the chip's overall performance. Higher array efficiency denotes longer BL and WL compromise including array access time [35,36].

4. Multi-level-cell phase change memory (MLC PCM)

As discussed in previous section, PCM refers to a memory technology helping in storage of data by programming and also read the resistance of the elements of the memory. Earlier research studies on PCM emphasized single bit operation but contemporary studies demonstrate the focus of MLC PCM operation, which keeps the low cost memory demand high [37]. Also, PCM devices can program the cell to arbitrary intermediate resistance values between the minimum SET resistance (10–50 KΩ) and the maximum RESET resistance (>1–5 MΩ).

It is possible to control the amorphous volume size within the cell by varying the pulse width, amplitude, or fall time of the electrical pulse [38]. The research studies further illustrated that the application of rectangular pulses with increase in amplitude result in proportionate increase in the cell resistance; and the cell resistance can consistently dropped by application of trapezoidal pulses with increase in the fall time [39,40]. These strategies can be helpful to demonstrate the impressive capability of MLC (up to 4 bits/cell) [41].

Fig. 3 highlights the MLC devices, which have storage capacity for 2 and 4 bits/cell, as the number of bits of every cell gets increased, resistance range of each level dropped exponentially. The highest number of bits in an MLC device is a function of reading precision, writing precision and device data integrity. The improvement in technology is met by the increase of bits in MLC devices. We suggest the use of 2-bit MLC model and projection of IRTS PIDS for devices of 3 and 4 bits which estimates Phase change memory's intermediate resistance value, energy profile and read/write latency.

Resistance partitioning of a 2-bit MLC PCM is shown in Fig. 4. It has $2\times$ gap of resistance among the adjacent states (i.e., $margin_{01-00}/margin_{10-01} = margin_{10-01}/margin_{11-10} = 2$). Fig. 4 shows the uniform partitions of GST

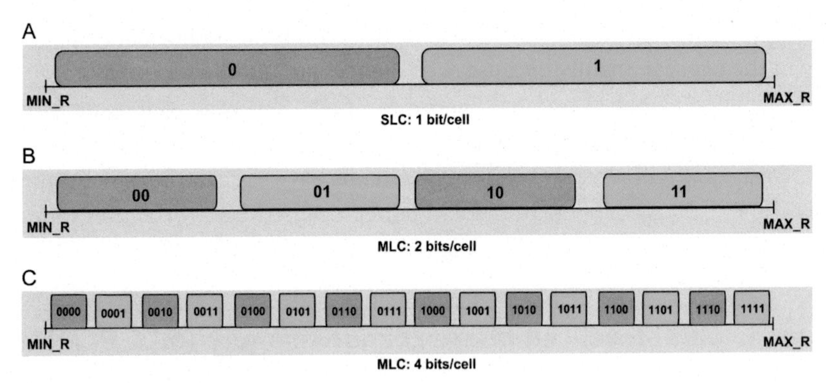

Fig. 3 Multi-level PCM concept.

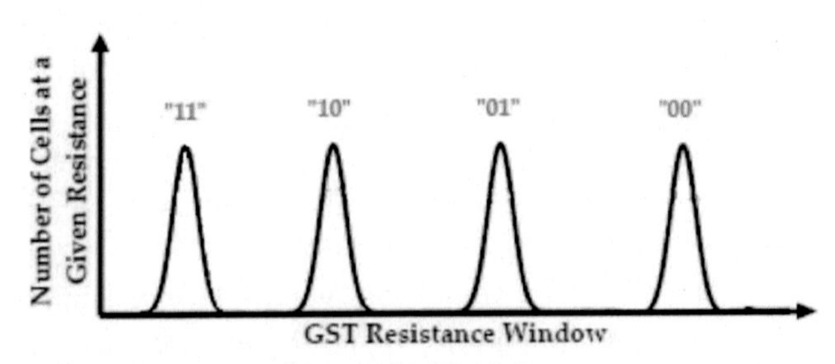

Fig. 4 Uniform resistance partitioning of 2-bit MLC.

resistance spectrum impacting on the access mechanism of read/write as well as cell reliability (we focus on the reliability aspect of MLC PCM in Section 7).

PCM cells can store 3 or 4 bits/cell in the future. To obtain tight resistance distributions in MLC PCM, iterative programming techniques are required [37]. By using a suitable write-read-verify programming schemes along with about 5–6 iterations in conventional PCM devices, we can properly control the distributions even for 4-bit MLC demonstrations [41]. In this line, it is necessary to manage the cell structure for programming to intermediate levels of resistance. For instance, incorporation of the multiple chalcogenide layers with distinguishing electrical resistivity [42] or cell structures with distinct parallel current paths [43].

5. Read techniques

Reading a PCM cell consists of sensing its level of resistance as well as mapping the analog resistance to an equivalent digital value. Read technique in SLC PCM consists of comparison the cell resistance to a reference cell (reference cell is the middle of the whole range of resistance).

For MLC PCM, read technique needs differentiating accurately between different levels of resistance placed in a closed manner. To read a data value, we apply a low amplitude as well as current pulse of short width to the GST material. By using a sample and hold (S/H) unit, the induced difference in voltage can then be captured [44]. To determine the bit-by-bit MLC cell's content, S/H output is fed into sequential analog-to-digital convertor (ADC) (for an N-bit MLC PCM, utilizing a minimum of 2^{N-1}). This approach, if followed, the bit-by-bit inspection of data value results in linear increase of energy and read latency with regard to the number of bits stored in an MLC cell.

6. Write techniques

In PCM, information can be stored in the cell in the form of n diverse levels of resistance signifying $\log_2 n$ bits since there is three orders of magnitude difference between the lowest and highest values of cell resistance. With increase in the number of levels of resistance stored in a cell, the resistance that is spread around the mean value tolerated by each level without mixing to the adjacent states of resistance reduces. In addition, relying on the value of resistance to be written/read, latencies of write and read are fluctuated. For MLC write, the GST's active portion is required to be partly crystallized or partly amorphized to program a cell to any kind of intermediate states. Also, the GST's amorphous fraction is required to be accurately controlled to achieve a required value of resistance within the predefined margin.

To program a MLC PCM cell, two approaches can be used, i.e., SET to RESET (S2R) and RESET to SET (R2S) programming [45,46]. In the first approach, the GST material's initial phase is presumed to be finally crystalline. Amorphous region is built by application of reset pulses of diverse amplitudes. The reset pulse exceeds the GST material's temperature above melting temperature leaving no time for crystallization because of rapid quench [47].

As illustrated in Fig. 5, this technique helps in causing amorphous and crystalline GST to be in series with each other. For placing the cell in diverse states of resistance, the amorphous cap's size should be controlled. Fig. 5 illustrated that amorphous cap with height h_2 has comparatively more voluminous than with height h_1. High resistivity amorphous material increases the resistance of the cell with amorphous cap of height h_2.

A second approach uses a fully reset state for the cell required to be programmed. Fig. 6 further illustrates that crystalline state can be built in the amorphous cap by application of the set current pulses. In order to modulate the crystalline volume, crystallization process is used. This in turn is necessary for placing the cell in intermediate resistance states by configuration of amorphous and crystalline GST.

For intermediate resistance values, sweep or staircase programming is used in which the highly amplitude initial pulse causes GST to melt. To reach an intermediate resistance state, discrete step and long sweep time or consistent decrease in the pulse's amplitude led material crystallization. To make sure proper formation of resistance, an iterative two-step

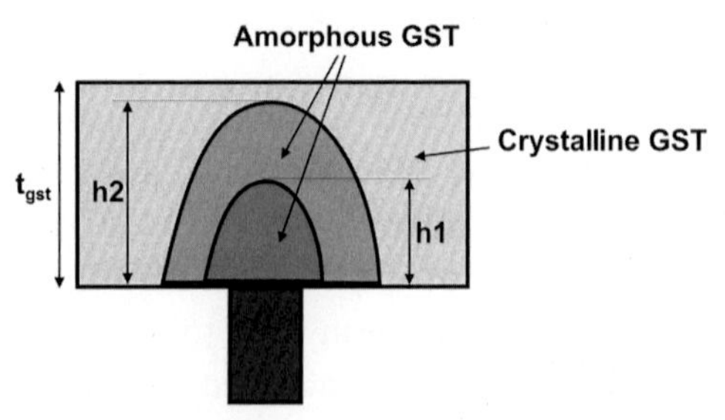

Fig. 5 Increasing amorphous region (h1 ⇔ resistance R1, h2 ⇔ resistance R2, h2 > h1 ⇒ R2 > R1).

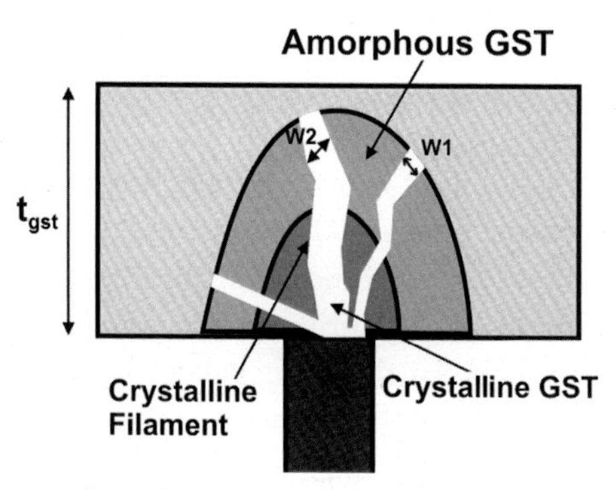

Fig. 6 Increasing crystalline filaments (w1 ⇔ resistance R1, w2 ⇔ resistance R2, w2 > w1 ⇒ R2 < R1).

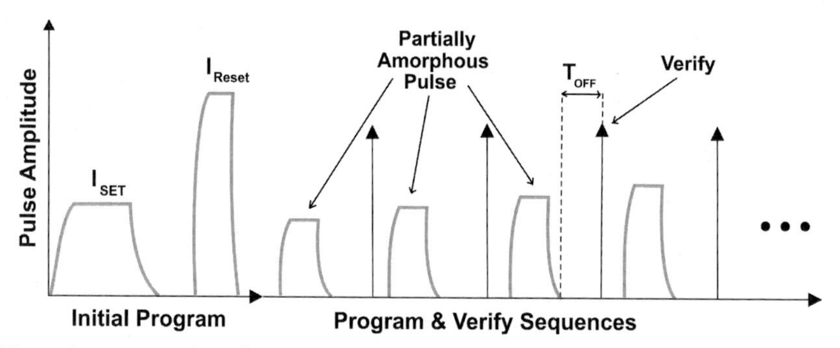

Fig. 7 Program and verify write process for MLC PCM.

Program-and-Verify (P&V) process is extensively exploited by some previous researches [37,48,49] (see Fig. 7).

The cell is required to experience a full SET pulse and a full RESET pulse as the initialization series which helps in improving the quality of programming. After this, a series of pulses with staircase up amplitude is implemented to the programmed cell based on the target resistance range, which is followed by a switch-off interval for relaxing the GST. Then, the read and verify process is done in order to check if the programmed resistance value falls within the target resistance range. If the resistance value does not, until the anticipated resistance is achieved, heuristic process is repeated.

Since iterative P&V process is used in MLC write process, initial RESET and subsequent staircase RESET pulses affect cell's lifetime in a negative manner. This scheme introduces the concerns related to reliability, which includes resistance drift.

7. Reliability

A basic understanding of failure mechanisms is necessary in order to have uniform performance from each of the billions of bits in the memory array. Current solutions propose some failure mechanisms for PCM [50–52].

Data retention is one of the primary reliability considerations of any non-volatile memory [53]. In this line, PCM may loss its data with two diverse mechanisms (Fig. 8). The first mechanism is related to the structural relaxation [54] (Fig. 8B) and thermally activated atomic reorganizations in the amorphous volume can increase the programmed level of resistance with time [55,56]. However, a reduction in the amorphous volume size due to crystallization can decrease the resistance with time [57] (Fig. 8C).

Therefore, the meta-stable nature of the amorphous state causes the resistance drift that is a source of multi-bit soft error for MLCs [58,59]. SLC PCM has the retention target of 10 years at 85 °C [34], so it is not susceptible to the resistance drift. After the succeeding sudden cooling of the cell, the cell's resistance grows for a particular time period before it again starts decreasing. On high resistance states, there is much more substantial drift, in which the GST's large volume is programmed to the amorphous, while the low resistance state reveals a nearly insignificant dependence of resistance on time.

The increase in the resistance rate follows a power-law behavior with the temperature and time. The model is defined as:

$$R = R_0 \left(\frac{t}{t0}\right)^{\nu}$$

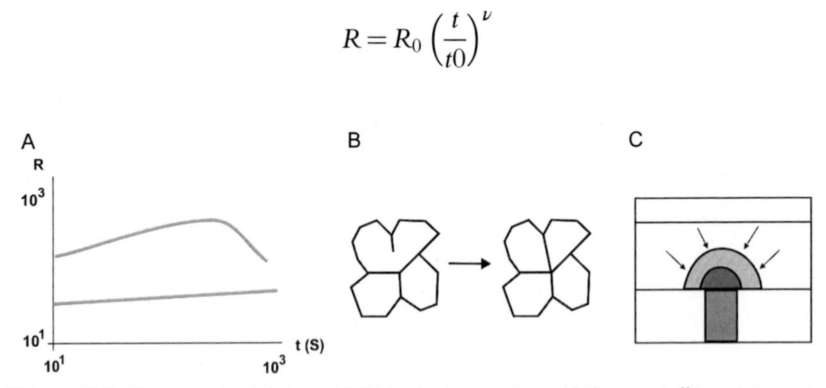

Fig. 8 (A) Drift upward initially, and then decreases toward the crystalline state resistance. The rate of drift is higher for larger resistances. (B) Structural relaxation, it shows the programmed resistance increases due to thermally activated atomic rearrangements in the amorphous volume and (C) crystallization state, it decreases as a result of the reduction in the size of the amorphous volume.

where ν is a drift coefficient which depends on the initial resistance level and temperature, its range is from 0 to 0.12. In that equation, R is denoted as GST resistance, after elapsed time of t with initial resistance of R_0 [60,61]. At a certain temperature, ν increases monotonically with the value of resistance following a logarithmic curve [61]. Temperature accelerates with ν for a certain resistance level and ν also increases monotonically with the resistance value.

Overlapping of the states in MLC led to wrong representation of bit values due to resistance shift of semi-amorphous states. To avoid regions' overlapping, we should determine the inter-state noise margins [62]. With an operating temperature of 300 K, margins with $2\times$ increase per step are estimated to be elongated enough based on the above power-law model (see Fig. 4). Alternative drift-tolerant approaches use ECCs, drift-aware coding [61,63–65], or pseudo refreshing Schemes [66] may impose energy and lifetime overheads [55,67].

Unlike SRAM or DRAM, PCM technology does not suffer from soft errors induced by charge-based radiation effects, which makes it appropriate even for space applications [68].

References

[1] Ovshinsky SR: Reversible electrical switching phenomena in disordered structures, *Phys Rev Lett* 21(20):1450–1453, 1968.
[2] Ovshinsky S, Fritzsche H: Reversible structural transformations in amorphous semiconductors for memory and logic, *Metall Trans* 17(2):641–645, 1971.
[3] Neale RG, Nelson DL, Moore GE: Non-volatile, re-programmable, readmostly memory is here, *IEEE Trans Electron* 56–70, 1970.
[4] Neale RG, Aseltine JA: The application of amorphous materials to computer memories, *IEEE Trans Electron Devices* 20(2):195–205, 1973.
[5] Pohm A, Sie C, Uttecht R, Kao V, Agrawal O: Chalcogenide glass bistable resistivity (ovonic) memories, *Proc IEEE Trans Magn* 6(3):1–17, 1970.
[6] Shanks R, Davis C: A 1024-bit nonvolatile 15ns bipolar read-write memory. In *Proceedings of the IEEE International Solid-State Circuits Conference,* vol. 3; 1978, pp 112–123, no. 7.
[7] Lai S, Lowrey T: OUM—a 180 nm nonvolatile memory cell element technology for stand alone and embedded applications. In *International Electron Devices Meeting*, 2001, pp 36–54.
[8] Burr GW, et al: Phase change memory technology, *J Sci Technol* 2(10):1–28, 2010.
[9] Wong HP, et al: Phase change memory, *Proc IEEE Trans Comput* 98(12):2201–2227, 2010.
[10] Friedrich I, et al: Structural transformations of $Ge_2Sb_2Te_5$ films studied by electrical resistance measurements, *J Appl Phys* 87(9):4130–4134, 2000.
[11] Gopalakrishnan K, et al: Highly-scalable novel access device based on mixed ionic electronic conduction (MIEC) materials for high density phase change memory (PCM) arrays. In *Proceedings of the International Symposium on VLSI Technology*; 2010, pp 205–217.

[12] Redaelli A, et al: Threshold switching and phase transition numerical models for phase change memory simulations, *J Appl Phys* 103(11):111101, 2008.
[13] Adler D, Henisch HK, Mott N: The mechanism of threshold switching in amorphous alloys, *Rev Mod Phys* 50(2):209–220, 1978.
[14] Jang MH, et al: Structural stability and phase-change characteristics of $Ge_2Sb_2Te_5/SiO_2$ Nano-multilayered films, *Electrochem Solid St* 12(4):34–38, 2009.
[15] Raoux S, et al: Direct observation of amorphous to crystalline phase transitions in nano particle arrays of phase change materials, *J Appl Phys* 102(9):18–23, 2007.
[16] Krebs D, et al: Threshold field of phase change memory materials measured using phase change bridge devices, *Appl Phys Lett* 95(8):1–18, 2009.
[17] Zhou G, Herman J, Borg JC, Rijpers N, Lankhorst M: Crystallization behavior of phase change materials: comparison between nucleation-and growth-dominated crystallization. In *Proceedings of the IEEE Optical Data Storage*; 2000, pp 74–76.
[18] Privitera S, Bongiorno C, Rimini E, Zonca R: Crystal nucleation and growth processes in $Ge_2Sb_2Te_5$, *Appl Phys Lett* 84(22):4448–4450, 2004.
[19] van Pieterson L, et al: Te-free, sb-based phase-change materials for high-speed rewritable optical recording, *Appl Phys Lett* 83(7):1373–1375, 2003.
[20] Chen YC, et al: Ultra-thin phase-change bridge memory device using GeSb. In *International Electron Devices Meeting*; 2006, pp 1–4.
[21] Rajendran B, et al: On the dynamic resistance and reliability of phase change memory. In *Proceedings of the IEEE Symposium on VLSI Technology*; 2008, pp 96–107.
[22] Sarkar J, Gleixner B: Evolution of phase change memory characteristics with operating cycles: electrical characterization and physical modeling, *Appl Phys Lett* 282–294, 2007.
[23] ITRS: *International Technology Roadmap for Semiconductors, ITRS.*
[24] Goux L, et al: Degradation of the reset switching during endurance testing of a phase-change line cell, *IEEE Trans Electron Devices* 56(2):354–358, 2009.
[25] Ohta T: Phase change optical memory promotes the DVD optical disk, *J Optoelectron Adv Mater* 3(3):609–626, 2001.
[26] Breitwisch M: Novel lithography-independent pore phase change memory. In *IEEE Symposium on VLSI Technology*; 2007, pp 100–104.
[27] Rajendran B, et al: Dynamic resistance—a metric for variability characterization of phase-change memory, *IEEE Electron Device Lett* 30(2):126–129, 2009.
[28] Atwood G, Bez R: 90nm phase change technology with ƒÊTrench and lance cell elements. In *Proceedings of the International Symposium on VLSI Technology, Systems and Applications*; 2007, pp 378–380.
[29] Im DH, et al: A unified 7.5nm dash-type confined cell for high performance PRAM device. In *IEEE International Electron Devices Meeting*; 2008, pp 1–4.
[30] Lee SH, et al: Programming disturbance and cell scaling in phase change memory: for up to 16nm based $4F^2$ cell. In *IEEE International Symposium on VLSI Technology*; 2010, pp 199–211.
[31] Kang S, et al: A 0.1 ƒÊm 1.8V 256-Mb phase-change random access memory (PRAM) with 66-MHz synchronous burst-read operation, *IEEE J Solid State Circuits* 42(1): 210–218, 2007.
[32] Lin L, et al: Driving device comparison for phase-change memory, *IEEE Trans Electron Devices* 58(3):664–671, 2011.
[33] Oh JH, et al: Full integration of highly manufacturable 512mb pram based on 90nm technology. In *IEEE International Electron Devices Meeting*; 2006, pp 1–4.
[34] Servalli G, et al: A 45nm generation phase change memory technology. In *Proceedings of the IEEE International Conference on Electron Devices Meeting (IEDM' 09)*; 2009, pp 1–4.
[35] Kang DH, et al: Two-bit cell operation in diode-switch phase change memory cells with 90nm technology. In *IEEE International Symposium on VLSI Technology*; 2008, pp 98–115.

[36] Lee K, et al: A 90nm 1.8V 512Mb diode-switch PRAM with 266MB/s read throughput. In *Proceedings of the IEEE International Conference on Solid-State Circuits*; 2007, pp 427–616.
[37] Bedeschi F, et al: A bipolar-selected phase change memory featuring multi-level cell storage, *IEEE J Solid State Circuits* 44(1):217–227, 2009.
[38] Hwang YN, et al: MLC PRAM with SLC write-speed and robust read scheme. In *IEEE International Symposium on VLSI Technology (VLSIT)*; 2010, pp 201–208.
[39] Lin J-T, Liao Y-B, Chiang M-H, Hsu W-C: Operation of multilevel phase change memory using various programming techniques. In *IEEE International Conference on IC Design and Technology*; 2009, pp 199–202.
[40] Nakayama K, MTakata T, Kasai AK, Akita J: Pulse number control of electrical resistance for multi-level storage based on phase change, *J Phys D Appl Phys* 40(17):55–67, 2007.
[41] Nirschl T, et al: Write strategies for 2 and 4-bit multi-level phase-change memory. In *Proceedings of IEEE International Electron Devices Meeting*; 2007, pp 461–464.
[42] Rao F, et al: Multilevel data storage characteristics of phase change memory cell with doublelayer chalcogenide films ($Ge_2Sb_2Te_5$ and Sb_2Te_3), *J Appl Phys* 46(2):25–37, 2007.
[43] Oh GH, et al: Parallel multi-confined (PMC) cell technology for high density MLC PRAM. In *Proceedings of the IEEE Symposium on VLSI Technology*; 2009, pp 220–232.
[44] Cabrini A, et al: Voltage-driven multilevel programming in phase change memories. In *IEEE International Workshop on MemoryTechnology, Design and Testing*; 2009.
[45] Kesltr W, editor: *Data Conversion Handbook, Analog Devices Series*, 2005, Elsevier.
[46] Razavi B: *Principles of Data Conversion System Design,* New York, 1995, Wiley-IEEE Press.
[47] Happ T, et al: Novel one-mask self-heating pillar phase change memory. In *International Symposium on VLSI Technology*; 2006.
[48] Joshi M, et al: Mercury: a fast and energy-efficient multi-level cell based phase change memory system. In *Proceedings of 17th International Conference on High-Performance Computer Architecture (HPCA-17)*; 2011.
[49] Qureshi MK, et al: Morphable memory system: a robust architecture for exploiting multi-level phase change memories. In *Proceedings of the International Symposium on Computer Architecture (ISCA)*; 2010, pp 153–162.
[50] Chimenton A, Zambelli C, Olivo P, Pirovano A: Set of electrical characteristic parameters suitable for reliability analysis of multimegabit phase change memory arrays. In *Proceedings of the International Non-Volatile Semiconductor Memory Workshop on Memory Technology and Design*; 2008, pp 49–51.
[51] Kim K, Ahn SJ: Reliability investigations for manufacturable high density PRAM. In *Proceedings of the IEEE International Symposium on Reliability*; 2005, pp 157–162.
[52] Pirovano A, et al: Reliability study of phase-change nonvolatile memories, *IEEE Trans Device and Mater Reliab* 4(3):422–427, 2004.
[53] Lacaita AL, Ielmini D: Reliability issues and scaling projections for phase change non volatile memories. In *IEEE International Electron Devices Meeting*; 2007, pp 157–160.
[54] Ielmini D, Lavizzari S, Sharma D, Lacaita AL: Physical interpretation, modeling and impact on phase change memory (PCM) reliability of resistance drift due to chalcogenide structural relaxation. In *IEEE International Electron Devices Meeting*; 2007, pp 23–36.
[55] Xu W, Zhang T: Using time-aware memory sensing to address resistance drift issue in multi-level phase change memory. In *Proceedings of the IEEE Symposium on Quality Electronic Design (ISQED' 10)*; 2010, pp 356–361.
[56] Jalili M, Sarbazi-Azad H: A compression-based morphable PCM architecture for improving resistance drift tolerance. In *IEEE International Conference on Application-specific Systems, Architectures and Processors (ASAP)*; 2014, pp 232–239.

[57] Shih YH, et al: Mechanisms of retention loss in $Ge_2Sb_2Te_5$-based phase-change memory. In *IEDM 2008. Proceedings of the IEEE International Electron Devices Meeting*, pp 1–4.
[58] Zhou J, et al: Formation of large voids in the amorphous phase-change memory $Ge_2Sb_2Te_5$ alloy, *Phys Rev Lett* 24–102, 2009.
[59] Zhong M, Song ZT, Liu B, Wang LY, Feng SL: Switching reliability improvement of phase change memory with nanoscale damascene structure by $Ge_2Sb_2Te_5$ CMP process, *Electron Lett* 44(4):322–323, 2008.
[60] Redaelli A, et al: Numerical implementation of low field resistance drift for phase change memory simulations. In *Proceedings of Non-Volatile Semiconductor Memory Workshop*; 2008, pp 39–42.
[61] Zhang W, Li T: Helmet: a resistance drift resilient architecture for multi-level cell phase change memory system. In *Proceedings of the IEEE/IFIP International Conference on Dependable Systems and Networks (DSN)*; 2011, pp 197–208.
[62] Fantini P, et al: Characterization and modeling of low-frequency noise in PCM devices. In *IEEE International Electron Devices Meeting*; 2008.
[63] Close GF, et al: Device, circuit and system-level analysis of noise in multi-bit phase-change memory. In *International Electron Devices Meeting (IEDM)*; 2010, pp 51–54.
[64] Lastras-Montano LA, et al: An area and latency assessment for coding for memories with stuck cells. In *Proceedings of Globecom Workshops (GCWkshps)*; 2010, pp 1851–1855.
[65] Jalili M, Arjomand M, Sarbazi-Azad H: A reliable 3D MLC PCM architecture with resistance drift predictor. In *International Conference on Dependable Systems and Networks (DSN)*; 2014, pp 204–215.
[66] Awasthi M, et al: Efficient scrub mechanisms for error-prone emerging memories. In *Proceedings of the IEEE International Symposium on High Performance Computer Architecture (HPCA)*; 2012, pp 15–26.
[67] Awasthi M, et al: Handling PCM resistance drift with device, circuit, architecture, and system solutions. In *Proceedings of Non-Volatile Memory Workshop*; 2011, pp 23–35.
[68] Maimon JD, Hunt KK, Burcin L, Rodgers J: Chalcogenide memory arrays: characterization and radiation effects, *IEEE Trans Nucl Sci* 50(6):1878–1884, 2003.

CHAPTER THREE

Phase-change memory architectures

Marjan Asadinia[a], Hamid Sarbazi-Azad[b]

[a]University of Arkansas, Fayetteville, AR, United States

[b]Sharif University of Technology and Institute for Research in Fundamental Sciences (IPM), Tehran, Iran

Contents

Advances in Computers, Volume 118
ISSN 0065-2458
https://doi.org/10.1016/bs.adcom.2019.10.001

Abstract

Some of the recent approaches regarding leverage PCM will be reviewed in this chapter. The chapter starts with a discussion regarding future main memory systems that includes hybrid architecture schemes using both PCM and DRAM arrays. Later, we focus on PCM only approaches and this section will help describe some techniques for reducing the increased read latency because of slow writes in PCMs.

In this chapter, we also illustrate wear-leveling approaches and review the security problems of this memory approach which are lifetime limited. This section includes an overview of the recent security aware wear-leveling techniques, whose methods help detect attacks, and their issues during the runtime.

Finally, we describe efficient schemes with hard error detection and correction capabilities as well as soft error problems like resistance drift.

1. Architecting PCM for main memories

Architects and system designers try to adventure the new memory technologies, which are able to provide more memory capacity than DRAM, even while being competitive in terms of performance, power, and cost.

Flash and PCM are known to be the promising technologies, which are helpful in bridging the access latency gap between DRAM and hard disk. Different memory technologies, their relative place in the overall memory hierarchy and their typical access latency (in cycles, assuming a 4 GHz machine) are shown in Fig. 1. The figure shows the speed difference between DRAM and hard disk can be bridged by a specific technology denser than DRAM and its access latency is within DRAM and hard disk.

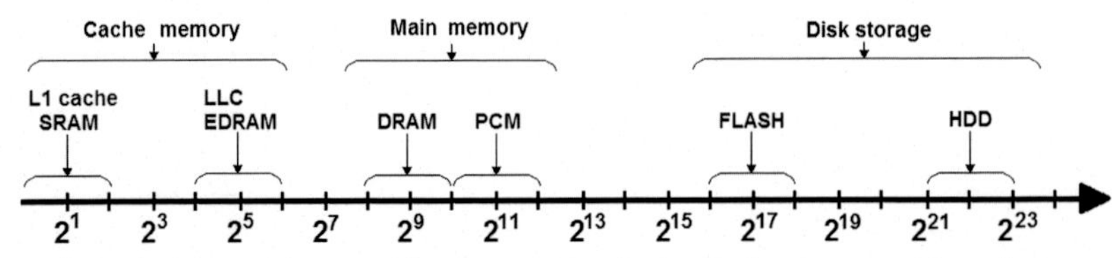

Fig. 1 Typical access latency of various technologies in the memory hierarchy.

To bridge the gap and reduce the power consumption of Hard Disk Drives (HDDs), Flash-based disk caches have been approached.

While compared to DRAM, Flash is two to three times slower but there is a need to raise the capacity of the DRAM for reducing the accesses to the Flash-based disk cache. Still DRAM has much closer access latency to the PCM and if we add the PCM density advantage, then there is no doubt PCM can attract much attention as an appealing technology which can increase memory capacity. Also, compared to the Flash cells, the cells of PCM can maintain 1000× more writes which creates the lifetime of PCM-based memory systems higher (in the range of years) in contrast to the range of days for a Flash-based memory system.

1.1 PCM organization

Array architectures as well as buffer organizations are presented in Lee et al. [1]. It was observed that as in the traditional memory organization, PCM cells could be organized into blocks, banks and even sub-blocks. The main thing is the benefit derived in terms of energy as well as performance when PCM specific optimizations are made to the architecture of memory array. It means that if the data in the row buffer has not been modified and PCM reads are non-destructive, it is not required to write back the row buffer.

Cross-coupled inverters are used to acquire both sensing and buffering mechanisms in conventional DRAM architectures, whereas, in PCM architectures sensing and buffering are separated which in turn enables multiple buffered rows to provide flexibility in row buffer organization. The separation of sensing and buffering mechanism enables multiplexed sense amplifiers, which drives banks of explicit latches and enables buffer width, a critical design parameter, much narrower than array width.

Reduction in area due to linearly deduction of the number of sense amplifiers with buffer width enables implementation of latched with multiple rows that are much smaller as compared to the removed sense amplifiers. Also, reduction in PCM write energy occurs due to narrow widths as each memory write is of a quite smaller granularity. This reduction can have negative impact on performance especially for high spatial locality applications, but the problem is alleviated by the presence of larger number of row buffers with higher buffer hit rate.

On comparing PCM array's delay and energy characteristics of diverse buffer design, Lee et al. [1] observed that decreases in both energy and delay was not the single 2KB buffer for DRAM but rather four 512KB wide buffers. This led to decrease the disadvantages of delay and energy of PCM from 1.6× and 2.2× to 1.1× and 1.0×, respectively, and illustrated that PCM is a tough competitor to DRAM.

1.2 Fine-grained write filtering

In PCMs, write process has high energy and can limit the systems' lifetime. The techniques that focus on decreasing the write traffic in PCM can also help to reduce the energy consumption along with increasing the lifetime of system. A traditional access to DRAM also updates a single write for the entire row, which is called as a page considering a memory bank. All the bits in the row are written only once. Though the higher portion of writes would be redundant, it doesn't apply where the write of a cell wouldn't change its value. It would be unnecessary to have these writes and taking them off could reduce the frequency of writes in corresponding cells as well. Fig. 2 demonstrates the bit writes percentage, which would be redundant for various benchmarks [2].

In this figure, authors measure the percentage of redundant bit writes above the total number of bits in write accesses. SLC series would denote a redundant bit write where each cell would store 1 or 0 in a single level cell. However, MLC-4 and MLC-2 represent a PCM cells of 4- and 2-bit width. Read operations are more critical than writes and increasing the latency of write has less negative impact on performance of the system. Fig. 3 shows the comparison logic of this approach and would be implanted through addition of XNOR gate in the cell's write path. PMOS gets connected to XNOR output, which has the potential to block the write current when the data of write equals to the present data that is stored. Gate of

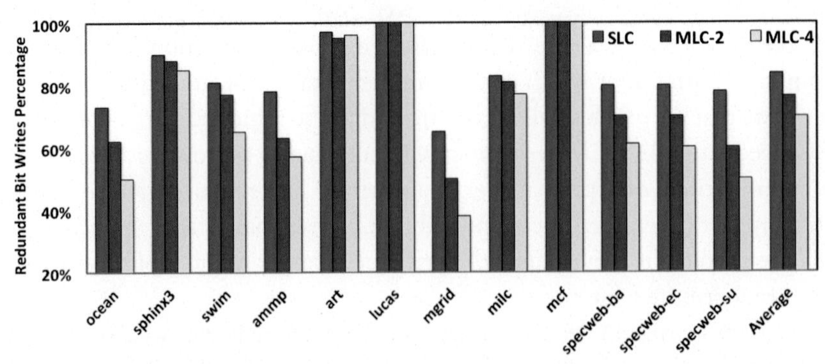

Fig. 2 Percentage of redundant bit writes for single-level and multi-level PCM cells.

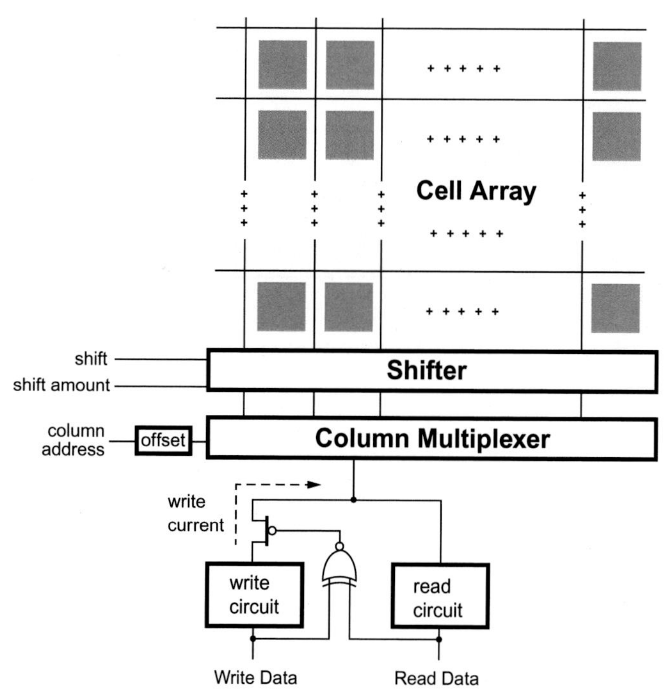

Fig. 3 Circuitry for redundant bit removal and row shifting.

XNOR is designed based on logic of pass transistor which has a simple structure producing small delay as well as less power consumption. There is spatial correlation in frequently written bits and Zhou et al. [2] had suggested *row shifting* scheme in order to have uniform write traffic across the bits of line.

1.3 Hybrid memory: Combining DRAM and PCM

To increase main memory capacity, PCM should be a good replacement of DRAM. However, the comparatively higher write latency in PCM as contrast to DRAM has a potential to degrade the performance of such a system. Therefore, *Hybrid Memory* that includes PCM-based main memory and a DRAM buffer is proposed in [3]. Such design combines the DRAM latency benefits as with PCM capacity benefits (Fig. 4).

The larger the PCM storage, the higher the capacity it provides to store pages during executing a program that reduces the disk access. DRAM memory, which is fast, would act as a buffer for the main memory as well as a medium between the main memory of PCM and processor. The access stream's locality solves the latency gap between PCM and DRAM.

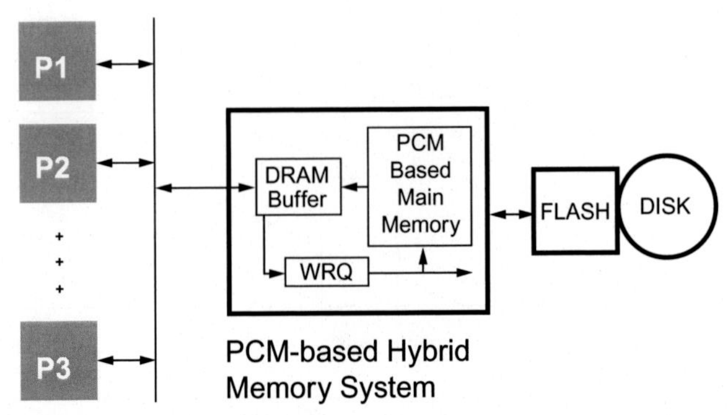

Fig. 4 PCM-based hybrid memory system.

In architecture having hybrid main memory, the Operating System which uses a page table can manage the PCM storage same as the present DRAM main memory systems. DRAM controller also manages the DRAM buffer. Organization of PCM storage along with DRAM buffer is at page granularity. When dirty pages are evicted from DRAM buffer, write traffic is visible to PCM. Since PCM has high write latency, WRQ or write queue would be given to buffer pages which are dirty and remove them from buffer of DRAM to be written in PCM.

2. Tolerating slow writes in PCM

PCM has high write latency and if there is sufficient write bandwidth, it uses buffers and intelligent scheduling to tolerate it. However, once a write request comes from PCM bank, access to the same bank for arriving read request for a different line should be postponed until the write request gets completed.

The write requests impact on the read request latency and cause latency to increase. Read is then in the critical path, and slowing down the read accesses, leads to performance loss. To solve this problem, this section addresses the slowdown issue as a result of long write latency and then alleviates the performance loss by introducing simple extensions to a PCM controller [4].

2.1 Write cancellation for PCM

In order to process the pending read requests when write request contention happens, a simple method for effective read latency is proposed in [5] which tries to cancel the write. This method is named as "*Write Cancellation*" and implemented in PCM devices with the ability to stop the write in case the write signal is currently not activated. Since the data placed at the address being written is in a non-deterministic state, they keep the copy of correct data in a write queue until the completion of write process. This method can reduce the read latency from 2290 to 1486 cycles.

However, to stop the deficiency of writes, the operation of *write cancellation* would be executed when the WRQ is lesser than 80%. But heavy read traffic leads to request of write being used when WRQ would exceed 80%. However, when it falls below the threshold of 80%, the read priority as well as *write cancellation* process would be enabled. These writes are result of WRQ occupancy when it is more than 80%, which is also called *forced writes*. These forced writes would be for less than 0.1% when compared with total writes of the baseline. But it can however increase to 2% of sum total of writes when the cancellation of write has to be used. As writes get priority over read requests, forced writes are limitation to the performance efficiency and threshold-based *write cancellation* should control them.

2.2 Write pausing

Writing, checking the cell's resistance value and rewriting are the steps of iterative writing. When iteration ends, this process can be stopped and serviced a pending read request for other line but in the same bank. This is called *write pausing* [5] and has a potential to improve read latency even if writes has many iterations while the end of each iteration denotes a pause.

Example of an iterative write with four iterations is shown in Fig. 5A. Three points at which pausing can happen is indicated in the figure. Assume that a read request arrives during Iteration 2 of the write, which is shown in Fig. 5B. At the end of Iteration 2, this request can be serviced with *write pausing*. Once the read request completes its service, Iteration 3 of the iterative write starts. Then, at pause points, Write pausing allows the read to be done transparently.

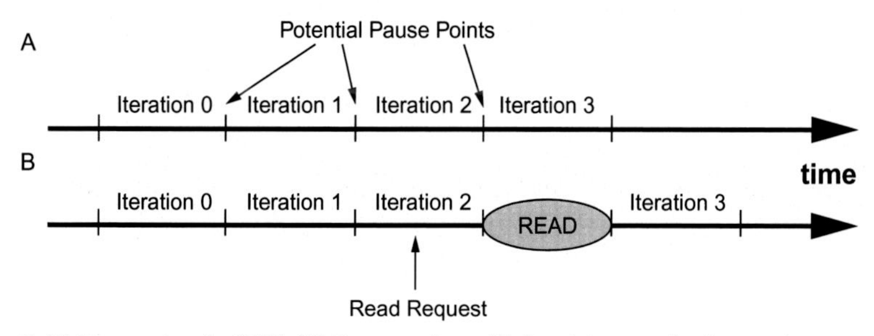

Fig. 5 Write pausing in PCM: (A) Pause points. (B) Servicing reads via pausing.

2.3 PRES: Pseudo-random encoding scheme to increase the bit flip reduction in the memory

Pseudo-Random Encoding Scheme (PRES) [6] is proposed to minimize the number of bit changes during memory writes. PRES uses the benefits of the advantage of mapping the write data vector into an intermediate highly random set of data vectors. When compared to the currently stored data, the intermediate data vector that yields the least number of differences is selected. Evaluation results show that PRES reduces bit flips by up to 25% over a baseline differential writing scheme. Further, PRES reduces bit flips by 15% over the leading bit flip minimization scheme, while decreasing encoding and decoding complexities by more than 90%.

3. Wear-leveling for durability

Wear-leveling techniques can be used to spread writes uniform over memory space. In this line, frequently written lines would be remapped to less frequently written lines. Existing proposals can use storage tables to track the write counts per line, some of these wear-leveling algorithms are presented in [7–11].

In wear-leveling schemes, logical lines are mapped to physical lines, however, this mapping is periodically changed and a separate indirection table is used for the storage of mapping. Table based wear-leveling methods would need as much megabytes hardware overhead as possible and this has a negative impact on the latency since at each memory access, indirection table must be searched for getting the physical location of a particular line.

To avoid the storage and latency overhead of existing methods and have the lifetime close to perfect wear-leveling, their proposals are presented in the following section.

3.1 Start-Gap wear-leveling

In table based wear-leveling methods, if an algebraic mapping of logical address to physical address is used, the storage and latency overhead of the indirection table can be resolved. In this line, Qureshi et al. [12] had suggested a Start-Gap wear-leveling which would use an algebraic mapping between both the physical and logical addresses avoiding the problem of tracking the write counts per line.

This Start-Gap would be efficient for wear-leveling as it moves periodically each line to a near location without considering the lines' write traffic.

This creates two types of registers: *Start* and *Gap* along with an additional memory line, *Gap Line*, to help the movement of data. The number of lines relocated in memory can be tracked by *Gap*. To track of how many times all the lines in memory have been relocated is kept by *Start*.

Since *Gap* and *Start* registers are changed continuously, logical to physical memory addresses mapping is also changed. Mapping is done by considering two observations: (1) All addresses less than *Gap* are remained unchanged, and all addresses greater than or equal to *Gap* are moved by 1 as it is shown in Fig. 6C. (2) As Fig. 6E shows, all locations move by 1 when *Start* moves, so to obtain the physical address, the value of *Start* must be added to the logical address. Compared to the baseline, Start-Gap can improve endurance by 10×. This method also needs storage for *Start* and *Gap* registers, each less than four bytes and incurs a total storage overhead of less than eight bytes for the entire memory.

3.2 Randomized Start-Gap

The design of Randomized Start-Gap algorithm is demonstrated in Fig. 7 [12]. Randomizer does a random mapping of a particular logical address to an intermediate address.

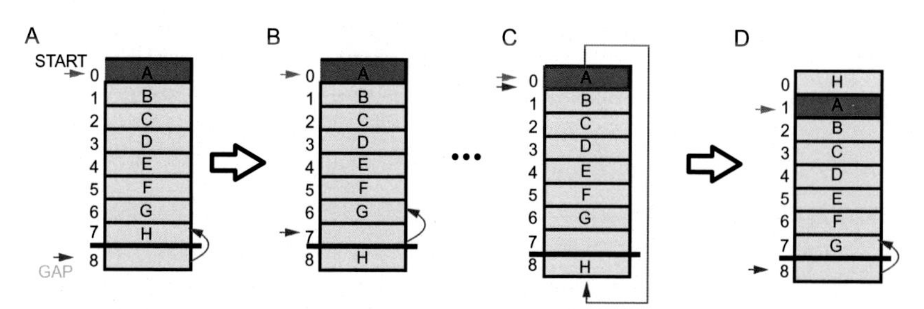

Fig. 6 Start-Gap wear-leveling on a memory containing eight lines.

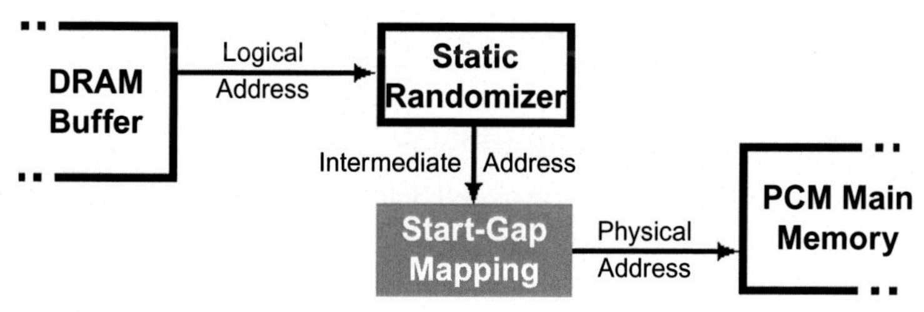

Fig. 7 Architecture for randomized Start-Gap.

Since IA and LA are randomly assigned, all regions have write traffics similar to the average. In this line, spatial correlation in the written lines that are heavy in LA doesn't have any impact on written lines between IA. As described in Fig. 7, logical address (LA) is the address produced from translation by OS and physical address (PA) denotes the physical location.

Each line in IA is mapped to the corresponding line in LA by the randomizer. The randomizer function is invertible. When static randomizer is used during program execution, it can help to store the randomized mapping constant and prevent from remapping. Therefore, the implementation of randomizing logic is possible only when the latency and hardware overhead is minimum.

By using an Address-Space Randomization method based on Feistel Network or a Random Invertible Binary Matrix, the translation of a physical address into an intermediate address is performed.

Non-uniform write traffic in various memory lines has a negative impact on PCM lifetime. To make the writes uniform, wear-leveling can be helpful. However, some of these wear-leveling techniques use large tables to keep the mapping information of logical addresses to the physical one, which imposes large area and large latency overhead, so it would not be suitable for use in main memories.

4. Secure wear-leveling algorithms

Row shifting [2] and segment swapping [2] are proposed as the wear-leveling algorithms using the fine-grained write behavior of PCM. Also, these algorithms work great for typical workloads. Since these are very deterministic, the mapped line location can be found clearly. These are also vulnerable to the attacks, which repeatedly try to be written on same line [12–14]. However, these attacks impact on system lifetime and causing to reduce it even though they can implement quite simple. To provide the security for wear-leveling algorithms, remapping should be performed in a randomized manner although this way is difficult to guess the physical location of a particular line for adversary when the relocation is done.

4.1 Region-based Start-Gap (RBSG)

An overview of RBSG or Region-based Start-Gap is shown in Fig. 8A [12]. This approach contains multiple regions each managed by its own Start and Gap registers, so it is similar to the Start-Gap approach. We adjusted the RBSG region size in such a way that maintaining the low rate of one line

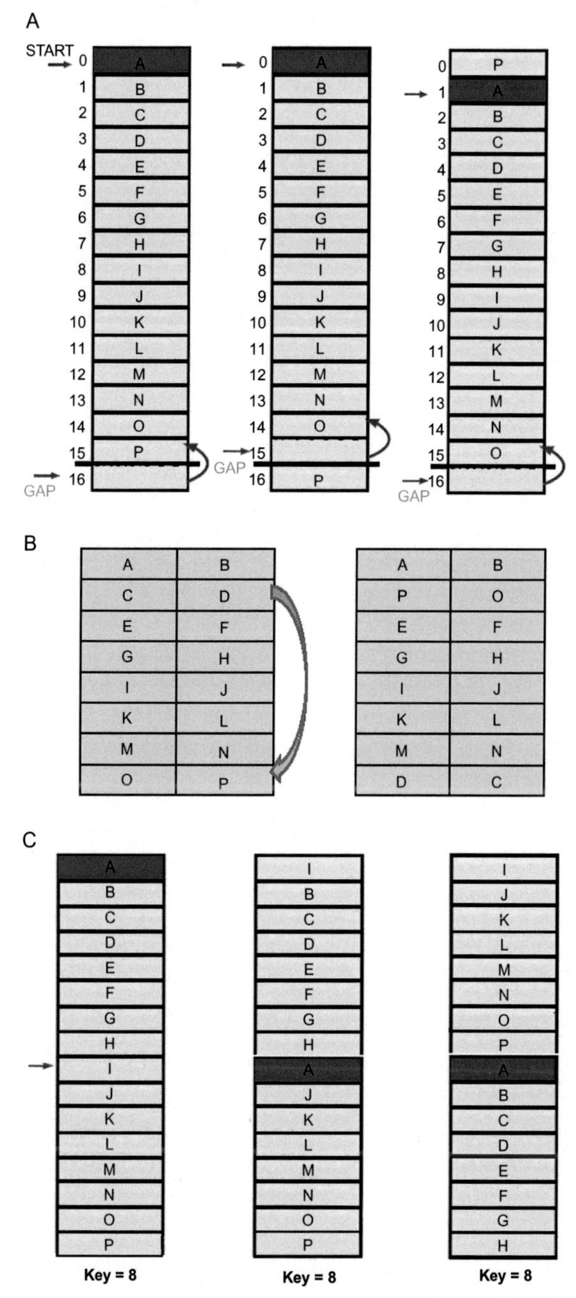

Fig. 8 Recent wear-leveling algorithms: (A) Region-based Start-Gap. (B) Seznec's PCM-S scheme. (C) Security refresh.

movement; the maximum number of writes to a line would be quite less than endurance. However, a movement having a low rate is simply not enough during attacks. It is shown in the study [13] that RBSG lifetime can last only for few days under the malicious write patterns with a movement having a low rate. Under attacks, robustness could be ensured through increasing the rate to greater than one line movement for every demand write.

4.2 PCM-S scheme

PCM-S (shown in Fig. 8B) was proposed by Seznec [15] and it is a scheme of table-based secure wear-leveling. For the table size to be reduced, the memory is however split into regions containing lines ranging in several thousands. Region is also swapped with a small probability *P* and a randomly picked region in memory. As it is shown by line D, after swaption the rotation of lines in the region are followed by a random amount. However, between the write overhead as well as the size of region, a trade-off exists and shows whatever is acceptable is the write overhead of one extra write per eight demand writes. Indeed, 12.5% write overhead is required in the scheme for every application, which isn't distributed gently over time, but is distributed in busts which means that two megabyte region sizes should get swapped. To stop large buffering, this scheme suggests region swaps with a sufficient rate and based on the demand writes. This consumes a quarter of overall write bandwidth, which is very effective.

4.3 Security refresh scheme

The security refresh scheme [14] is described in Fig. 8C. To determine the amount for memory regions to be rotated entirely, a random key is used. Two lines are swapped to perform the rotation. Once all lines get swapped and rotated, whole memory gets shifted through the amount defined by a random key and then the change takes place in the key value.

SR-1 or Single Level Security Refresh is robust but however its robustness could be improved through the memory being splitted into different regions. Remapping of each region is done within the region at a particular IR or inner rate ensuring the regions to be swapped at OR Outer rate. This describes SR-M or *Multi-Level-Security-Refresh* scheme. SR-M has a write overhead which could be reduced through IR, OR and controlling both of them. Since the write overhead of SR-M is a function of endurance, one way this write overhead can be reduced is the adaptation (using adaptive SR-M).

It is notable that the PCM-based memory system remains durable even under the most stressful write pressure when a robust wear-leveling algorithm is used. Recently suggested secure wear-leveling scheme, those robust towards attacks, provide memory remapping of memory at a greater rate when memory is assumed always under attack. High write overhead is then incurred and leading it ineffective for all practical purposes. Also, it is desired to have a low write overhead as well to add robustness to the attacks.

4.4 SLC-enabled wear-leveling for MLC PCM

According to [16], authors use adaptive and dynamic mode transformation of SCL from MLC to present SLC-enabled wear-leveling scheme. To achieve remarkable endurance benefits, instead of redistributing write operations, this proposed method dynamically transforms weak and write dense components into SLC mode.

This scheme entails an additional memory space that always works in SLC mode and helps MLC pages to transfer into SLC mode. Indeed, when one MLC page is chosen for transfer, with the aid of extra memory space, the content of the selected MLC page is rewritten into SLC format. In this line, transformation is performed via the rewriting of the initial MLC page and the extra SLC page. Each additional SLC page generally maintains a link, which indicates their corresponding MLC pages, so their content spreads to the current extra pages in an SLC-like manner.

5. Error resilience in phase change memories

When PCM is used as primary memory, many noteworthy impediments can arise. One of the most significant ones is the limited write endurance estimated to be currently in the range of few tens of millions. By considering this limited endurance, with the help of perfect and effective wear-leveling which guarantees that all cells inside the memory array can have a range of write endurance, it is possible to come up with a memory system surviving for several years. However, the downside of it all is the process variation has a potential of causing a significant variability in the various cells lifetime found in a similar PCM array. So, the overall system lifetime can be limited just by the cells which have endurance much lower than an average cells (called weak cells). For long lifetime achievement in such designs, there is a need for incorporation of two types of methods. Indeed, a large number of errors for each memory line or page should be either tolerated with graceful degradation or corrected with coding techniques.

5.1 Fault model assumption

Since PCM needs high energy to change its states, transient faults induced by radiation are not able to influence on PCM cells. Permanent failure of a cell is the major fault model to be applicable to the cells of PCM. Writing a cell requires heating and cooling procedure and also would result in contraction and expansion. It will lead to the element of heat to detach itself from the Chalcogenide material proving impervious for programming pulse. This write induced faults are captured as a struck-at-fault. Fortunately, by using the popular stuck-at-fault detection mechanism, which is read-after-write scheme, the content of such stuck cells can be read reliably [17]. It is notable there is no large systematic effect within a die, and in general in presence of such permanent faults. Using additional circuitry can compensate and trim such positional effects [17].

5.2 Dynamically replicated memory (DRM)

The correction schemes at fine granularity (like ECP or PAYG) permanently disable a block when it encounters the first uncorrectable error, which may result in a sudden reduction in PCM capacity after a few years. This has been the motivation for proposing DRM [17] using an architectural OS-support technique for further improving PCM lifetime. In this scheme, each byte has its own failure indication bit. When a new failure occurs, the indication bit of the corresponding byte is set, and the OS adds the corresponding page to a waiting list of faulty pages. Upon page allocation, OS selects a pair of faulty pages such that their bit failures are not at the same offset within the page. At this point, one of these two pages is maintained as the primary copy and the other one is used as backup. Despite its simplicity, DRM makes the memory system vulnerable to malicious attacks, especially when the OS is compromised. Also, it does not support wear-leveling techniques manipulating memory block addresses to uniformly spread the writes over different blocks.

5.3 Error correcting pointers (ECP)

As an efficient alternative to ECC, ECP corrects errors by logging the bit-errors in a line [18]. For each bit failure, a pointer is used to refer to the position of the bit-error and an additional bit to hold the correct value. ECP can recover multiple bit failures irrespective of their positions. ECP's storage overhead depends on the recoverable number of bit-errors and memory line size. Evaluations show that ECP-6 (ECP handling up to 6 bit-errors) provides a lifetime of about 6.5 years with 12% storage overhead. ECP has, however, some major drawbacks: it can tolerate only wear-out failures, so augmenting ECP with ECC to achieve end-to-end reliability incurs unnecessarily large storage overhead. Moreover, the hard error tolerance strength of ECP is fixed at design time.

5.4 Stuck-at-fault error recovery (SAFER)

SAFER is proposed by Seong et al. [19] and uses dynamic partitioning technique in which ensures each partition has at most one failed error in presence of stuck-at faults. It relies on DEC or Double Error Correction and lifetime evaluation is shown that SAFER could have a better lifetime compared to the ECP-6 but it just imposes 55 bits overhead instead of the present 61 bits in ECP-6.

But the main problem of the SAFER is it needs extra writes for error correction to get performed which comes at the cost of additional power, performance loss and the additional penalties for wear-out. By utilizing a fail cache to identify the failed bit positions, extra write could be avoided. It will also lead the data to be morphed so as to mask the hard fault on the first write but it is not usable for multi-bit error correction since it allows maximum one failed bit. Also, it comes with extra design as well as area overhead.

5.5 Fine-grained embedded redirection (FREE-p)

FREE-p [20] is a protection scheme designed for tolerating both hard and soft errors (at access logic) in PCMs. For hard errors, FREE-p utilizes the still-functional cells within a faulty block and stores remapping information with high redundancy (8 × redundancy) for future references. In contrast to DRM, FREE-p does not use an extra structure for remapping, resulting in an almost zero storage overhead. For soft errors, on the other hand, a special ECC is designed for end-to-end detection and correction support. FREE-p exhibits a large latency overhead to access fine-grained remapping information. To mitigate this effect at architecture level, FREE-p relies on a caching mechanism to reduce multiple accesses to the PCM array. This structure, however, increases the complexity of the ultimate PCM design with respect to both hardware and OS requirements (Fig. 9).

5.6 A recursively defined invertible set scheme to tolerate multiple stuck-at faults in resistive memory (RDIS)

In case of stuck-at faults, the RDIS tries to retrieve the data correctly [21]. Indeed, during the read process, it tracks the bits stuck at various values

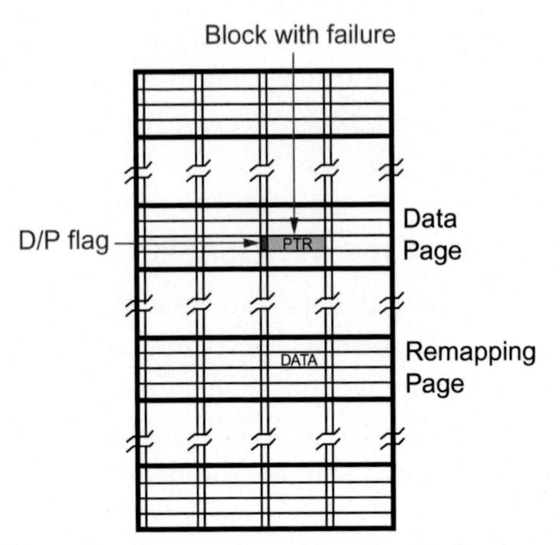

Fig. 9 Example of fine-grained remapping with embedded pointer.

differing from written values. We can get the correct values by inverting the values read for those bits likewise during a write operation, each cell in a data block has different types. They are either Non-Faulty (NF), stuck against the value which is written struck at wrong (or also called as SA-W), or struck at same written value which is also called as stuck at right (or also SA-R).

Assume we want to write 0 where the cell is stuck at 1, then it would make the cell SA-W. RDIS tries to identify and encode a subset S out of all cells in data block which has to get updated and contain every cells of SA-W. We then read the inverted form of S members which retrieve the originally written data. RDIS uses read-after-write verification operation for detecting the write failure and initiates S computation after its detection mechanism.

5.7 Pay-as-you-go: Low-overhead hard error correction for phase change memories (PAYG)

The PAYG error correction scheme [22] is based on the key observation that only a few PCM lines (less than 1%) experience more than one permanent fault and need a high level of protection for a typical application. Therefore, statically allocating a fixed number of pointers to a line (e.g., 6 pointers in ECP-6) wastes storage resources. PAYG proposes a two-level structure of correction pointers where each line corresponds to one dedicated ECP entry and the remaining entries are kept in a common pool to be used on demand by all lines. Evaluations show PAYG requires $3\times$ lower storage overhead than ECP-6 and yet provides 13% more lifetime while incurring a negligible latency overhead (less than 0.4%).

5.8 Zombie scheme

To extend PCM lifetime and tolerate more stuck at faults, Zombie [23] framework is proposed. It relied on pairing the primary blocks in working disabled memory pages and dead with other spare blocks, which adaptively provide error correction resources to primary blocks. Authors also mention that Zombie framework can be applied by combination of two error correction mechanisms (ZombieXOR for SLC and ZombieMLC for MLC) and the extension of two proposed SLC mechanisms (ZombieECP and ZombieERC).

5.9 Aegis method

As partitioning and inversion mechanism are used by some existing state-of-the-art solutions. Aegis [24] also has proposed to use bit partitioning and group inversion. The efficiency of this approach largely depends on how a data block is partitioned into groups. Aegis can re-partition a data block so that it guarantees any two bits in the same group will not be in the same group after a re-partitioning. Compared to the existing recovery mechanism, it can recover more faults with reduced space overhead. However, this scheme suffers from increasing bandwidth demand as a result of increasing the number of bit failures.

5.10 Captopril scheme

The problem of number of bit flips during write operation is addressed by Captopril [25]. This method used the read-before-write (RBW) approach masking the unchanged bits during write operation. Authors observed some specific locations of blocks are responsible for the majority of bit flips. So to solve the issue of long latency during write operation and non-destructive nature of reads in PCM, they used the read-before-write approach to further reduce the number of bit flips per write in the memory system.

5.11 Tolerating hard errors using compression

To prolong the lifetime of a MLC PCM main memory, authors in [26,27] proposed to postpone the occurrence of stuck-at failures by converting blocks from MLC to SLC mode using byte-level compression, and tolerate hard errors. They used some pointers and covered hard errors by fitting a compressed block either in its physical line location or elsewhere in the page. Compared to a state-of-the-art design, they reached 48% lifetime improvement of the memory system along with 9% IPC improvement.

5.12 Improving performance and lifetime with relaxed write/read

To correct permanent stuck-at hard errors, extra storage is needed per memory line. Since the extra storage is used after hard error occurrence, so its utilization is low for a long time before hard errors occurrence. To improve the read/write latency in a 2-bit MLC PCM, a relaxation scheme is proposed in [28] and called as Relaxed Write/Read (RWR) method to use the extra storage for reading and writing the cells for intermediate resistance levels. It is based on the programming of the intermediate state cells to the intermediate state of Tri-Level instead of their exact value in MLC mode. In fact, for improving write latency, this approach combined the most time-consuming levels (intermediate resistance levels) to reduce the number of resistance levels (making a Tri-Level PCM). To retrieve the exact data values in the read operation, some error correction metadata are stored in the extra storage section.

6. Soft error approaches

Recent studies have indicated that the dominant failure mode in PCM memory system is likely to be hard failure due to variance in lifetime of different cells of a PCM array. To tolerate such lifetime failures, this section described several techniques, including dynamic memory replication, write-efficient coding, and embedded redirection of failed lines. While the focus of past research has been dealing with endurance-related hard faults, PCM cells are also susceptible to newer failure modes due to resistance drift and read disturb.

Regarding resistance drift tolerance at architectural level, Zhang and Li [29] proposed Helmet, a suite of data encoding techniques, including rotation and inversion, to decrease the probability of storing data in drift-sensitive regions and therefore increasing reliability in a power-efficient manner. Awashti et al. [30] investigated the applicability of some pseudo scrubbing (refreshing) schemes to reduce the chance of encountering the second bit error when a single-bit error is found in a physical memory line. Wu et al. [31] advocated a time-aware design methodology to either dynamically determine inter-state noise margin boundaries or adaptively store encoded data in a reliability-enhanced PCM system.

Another work is presented in [32] and preserves the performance of NVMs while provides the reliability and availability for the large-scale storage systems. Indeed, by using a two-tier architecture in which the primary tier contains a mirrored pair of nodes and the secondary tier contains one or more secondary backup nodes with weakly consistent copies of data, this approach achieves these goals. In this line, it uses highly optimized replication protocols, software, and networking stacks to minimize replication costs and expose as much of NVM's performance as possible. This proposed method provides replicated NVMs with similar or even better performance than the other un-replicated NVMs (reducing latency by 27–63% and delivering between 0.4 and 2.7 × throughputs).

References

[1] Lee BC, et al: Architecting phase change memory as a scalable DRAM alternative. In *Proc. Int'l Symp. Computer Architecture (ISCA '09)*; 2009, pp 2–13.

[2] Zhou P, et al: A durable and energy efficient main memory using phase change memory technology. In *Proc. IEEE Symp. High Performance Computer Architecture (HPCA '09)*; 2009, pp 14–23.

[3] Qureshi MK, Srinivasan V, Rivers JA: Scalable high performance main memory system using phase-change memory technology. In *ISCA*; 2009, pp 24–33.

[4] Qureshi MK, Franceschini MM, Lastras-Monta LA: Improving read performance of phase change memories via write cancellation and write pausing. In *Proc. IEEE Symp. High Performance Computer Architecture (HPCA '10)*; 2010, pp 1–11.

[5] Xie Y, et al: A novel architecture of the 3D stacked Mram L2 cache for CMPs. In *Proc. IEEE Symp. High Performance Computer Architecture (HPCA' 09)*; 2009, pp 36–48.

[6] Seyedzadeh SM, Maddah R, Jones A, Melhem R: PRES: Pseudo-Random Encoding Scheme to increase the bit flip reduction in the memory. In *Proc. Design Automation Conference (DAC '15)*; 2015, pp 1–6.

[7] Kgil T, Roberts D, Mudge T: Improving NAND flash based disk caches. In *International Symposium on Computer Architecture (ISCA 08)*; 2008, pp 327–338.

[8] M-Systems: TrueFFSWear-Leveling Mechanism, 2002: http://www.dataio.com/pdf/NAND/MSystems/TrueFFS_Wear_Leveling_Mechanism.pdf.

[9] Ban A, Hasharon R: *Wear Leveling of Static Areas in Flash Memory,* U.S. Patent Number 6, 2004, pp 44–49.

[10] Ben-Aroya A, Toledo S: Competitive analysis of flash-memory algorithms. In *Proc. Annual European Symposium*; 2006, pp 100–111.

[11] Gal E, Toledo S: Algorithms and data structures for flash memories, *ACM Comput Surv* 37(2):138–163, 2005.

[12] Qureshi MK, et al: Enhancing lifetime and security of PCM based main memory with Start-Gap wear leveling. In *Proc. IEEE/ACM Int. Symp. Microarchitecture (MICRO '09)*; 2009, pp 14–23.

[13] Seznec A: *Towards Phase Change Memory as a Secure Main Memory,* Technical report, 2009, INRIA.

[14] Seong NH, Woo DH, Lee H-HS: Security refresh: prevent malicious wear-out and increase durability for phase-change memory with dynamically randomized address mapping. In *ISCA*; 2010.
[15] Seznec A: A phase change memory as a secure main memory. In *Proc. IEEE Computer Architecture Letters*; 2010, pp 57–62.
[16] Zhao M, Jiang L, Zhang Y, Xue CJ: SLC-enabled wear leveling for MLC PCM considering process variation. In *Proc. Annual Design Automation Conference (DAC)*; 2014, pp 1–6.
[17] Ipek E, Condit J, Nightingale EB, Burger D, Moscibroda T: Dynamically replicated memory: building reliable systems from nanoscale resistive memories. In *ASPLOS*; 2010, pp 3–14.
[18] Schechter SE, et al: Use ECP, not ECC, for hard failures in resistive memories. In *Proc. Int. Symp. Computer Architecture (ISCA)*; 2010, pp 141–152.
[19] Seong NH, et al: SAFER: stuck-at-fault error recovery for memories. In *MICRO*; 2010, pp 115–124.
[20] Yoon DH, et al: FREE-p: protecting non-volatile memory against both hard and soft errors. In *HPCA*; 2011, pp 466–477.
[21] Maddah R, Melhem R, Cho S: RDIS: tolerating many stuck-at faults in resistive memory, *IEEE Trans Comput* 64(3):847–861, 2015.
[22] Qureshi MK: Pay-as-you-go: low-overhead hard-error correction for phase change memories. In *Proc. IEEE/ACM Int. Symp. Microarchitecture (MICRO)*; 2011, pp 318–328.
[23] Azevedo R, Davis JD, Strauss K, Gopalan P, Manasse M, Yekhanin S: Zombie memory: extending memory lifetime by reviving dead blocks. In *Proc. Int. Symp. Computer Architecture (ISCA)*; 2013, pp 452–463.
[24] Fan J, Jiang S, Shu J, et al: Aegis: partitioning data block for efficient recovery of stuck-at-faults in phase change memory. In *Proc. the 46th Annual IEEE/ACM Int. Symp. Microarchitecture*; 2013, pp 433–444.
[25] Jalili M, Sarbazi-Azad H: Captopril: reducing the pressure of bit flips on hot locations in non-volatile main memories. In *Design Automation Test in Europe Conference Exhibition (DATE)*; 2016, pp 1116–1119.
[26] Jalili M, Sarbazi-Azad H: Tolerating more hard errors in MLC PCMs using compression. In *International Conference on Computer Design (ICCD)*; 2016, pp 117–123.
[27] Jalili M, Sarbazi-Azad H: Endurance-aware security enhancement in non-volatile memories using compression and selective encryption, *IEEE Trans Comput* 66(7):1132–1144, 2017.
[28] Rashidi S, Jalili M, Sarbazi-Azad H: Improving MLC PCM performance through relaxed write and read for intermediate resistance levels, *ACM Trans Archit Code Optim* 15(1):1–31, 2018.
[29] Zhang W, Li T: Helmet: a resistance drift resilient architecture for multi-level cell phase change memory system. In *Proc. IEEE/IFIP Int. Conf. Dependable Systems and Networks (DSN)*; 2011, pp 197–208.
[30] Awasthi M, et al: Efficient scrub mechanisms for error-prone emerging memories. In *Proc. IEEE Symp. High Performance Computer Architecture (HPCA)*; 2012, pp 15–26.
[31] Wu Q, et al: Using multi-level phase change memory to build data storage: a time-aware system design perspective, *IEEE Trans Comput* 62:2083–2095, 2013.
[32] Zhang Y, Yang J, Memaripour A, Swanson S: Mojim: a reliable and highly-available non-volatile memory system. In *Proc. Int. Conf. Architectural Support for Programming Languages and Operating Systems (ASPLOS)*; 2015, pp 3–18.

CHAPTER FOUR

Inter-line level schemes for handling hard errors in PCMs

Marjan Asadinia[a], Hamid Sarbazi-Azad[b]
[a]University of Arkansas, Fayetteville, AR, United States
[b]Sharif University of Technology and Institute for Research in Fundamental Sciences (IPM), Tehran, Iran

Contents

Abstract

To address the problem of fast degradation in PCM main memory systems in the presence of severe cell wear-out, this chapter introduces and evaluates some ways to deal with hard error issues in phase change memory. Our observation reveals when some

Advances in Computers, Volume 118
ISSN 0065-2458
https://doi.org/10.1016/bs.adcom.2019.10.002

memory pages reach their endurance limits, other pages may be far from their limits even when using a perfect wear-leveling. Recent studies have proposed redirection or correction schemes to alleviate this problem, but all suffer from poor throughput or latency.

In this chapter, we also propose On-demand page paired PCM (OD3P) memory system. Our technique mitigates the problem of fast failure of pages by redirecting them onto other healthy pages, leading to gradual memory capacity degradation. We then extend our proposed scheme at line-level, called line-level OD3P as a way to increase the durability of PCM pages by enabling line pairing within a page. Afterwards, we introduce our simulation environment and experimental results of our evaluations.

1. OD3P: On-demand page paired PCM

Studying process variation in PCM arrays shows there is no spatial correlation of failures among neighboring cells; it implies that in the absence of error recovery techniques, the weakest cells dictate the lifetime of a PCM memory block. Therefore, different memory blocks exhibit different levels of durability.

In a large PCM array, this variation impacts the wear-leveling efficiency and may cause sudden reduction in memory capacity. This fact, motivating our proposal, can be easily revealed as follows: assume a 1 GB PCM memory bank with 1 KB pages and 128 B lines as read/write granularity. To model process variation, we consider a model where the endurance of each individual cell is set by a random variable with a normal distribution of 10^8 writes (in average) and a standard deviation of 0.25×10^8 (coefficient of variation, CoV of 0.25 [1,2]). A perfect wear-leveling scheme is used to evenly distribute write accesses over the whole address range. A PCM line is assumed to tolerate up to six errors, beyond which that line (as well as its corresponding page) is marked as faulty. We measured this duration (when a line becomes faulty) and plotted it (normalized to the endurance of a page with the lowest lifetime) in Fig. 1.

It is seen that, even after applying wear-leveling, there are some pages that are far from their endurance limits, when some pages become faulty.

Our proposed method is described in the next section but we first want to discuss error detection and correction mechanisms used at lines of a page in OD3P architecture. In OD3P, the memory controller issues a read command after each write to determine whether the write is done successfully. The key observation that makes this approach practical is that PCM writes are much more expensive (in terms of latency and energy) compared to

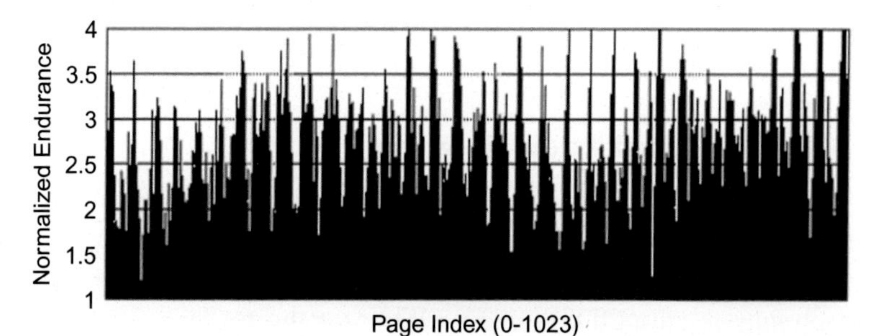

Fig. 1 Write endurance of different 1 KB pages in a 1 GB PCM bank. The average cell's endurance is 10^8 with standard deviation of 0.25×10^8. The values are normalized to the weakest page (3.8×10^7 in this experiment). For considerable number of pages, endurance varies by $4\times$ that cause sudden reduction in memory capacity even after wear-leveling.

PCM reads. Therefore, performing read-after-write for error detection, as read is much faster than write, has a negligible impact on overall system performance. For instance, our experiments confirm that this mechanism results in less than 0.4% IPC degradation of the baseline system in Section 7 (with PCM read and write latency of 50ns and 500 ns, respectively) when running different applications.

For error correction, we use ECP-6 for each memory line. When detection mechanism determines a write operation was not performed properly for a particular bit, the position of the bit error is recorded in a free ECP pointer associated with that line and the correct bit value is stored in its corresponding bit storage. Once all pointers of a line are allocated, any further bit failure detection translates to an uncorrectable error and marks the line as faulty. At this point, the memory controller invokes our recovery mechanism.

2. Structure and operation of page paired PCM

We observed, in Fig. 1, how process variation could affect PCM page wear-out: at a point of page failure, there are some pages that are far from their endurance limit (healthy pages). The faulty and healthy pages can be paired in MLC mode in order to improve the overall PCM lifetime and prevent sudden reduction of memory capacity. The operation of OD3P system involves a target page selection algorithm that determines the address of target page; a pairing mechanism that realizes the pairing process of the faulty and target pages, and an *address translation* mechanism that translates the OS generated physical addresses to PCM addresses.

2.1 Target page selection algorithm

This algorithm determines with which PCM page a given faulty page can be paired. We start with the approach that gives the maximum freedom for choosing the target page from all healthy memory pages (we call this approach *fully-selective* OD3P, or FS-OD3P for short).

It is intuitive that the target page should be the healthiest PCM page, which has not been under write stress recently. Since we use wear-leveling to uniformly distribute write accesses, the number of unallocated pointers in ECP can be a good indication of page health. When wear-leveling works properly, we can assume all pages receive almost even write traffic. Therefore, no special mechanism is required to track write stress of different pages.

It is clear that FS-OD3P imposes excessive latency to search through thousands of memory pages to find the most suitable target page. Instead, we propose a much simpler and faster selection mechanism (with an outcome comparable to FS-OD3P), called *Target Page Selector* (TPS).

Fig. 2 gives a structural view of the TPS. TPS has a table with few entries to keep the information of a limited number of most healthy pages. Each entry consists of a *valid* bit, which indicates whether the entry is valid, index of the PCM page, the entry is associated to *(Index)*, and the number of free ECP pointers of that page (*FreeECPs*). A tree structure of comparators and multiplexers allows a register *Max* be updated with the maximum value of FreeECPs pointers in TPS (*Max.FreeECPs*), as well as the position of the corresponding entry in TPS (*Max.Entry*), and corresponding page address

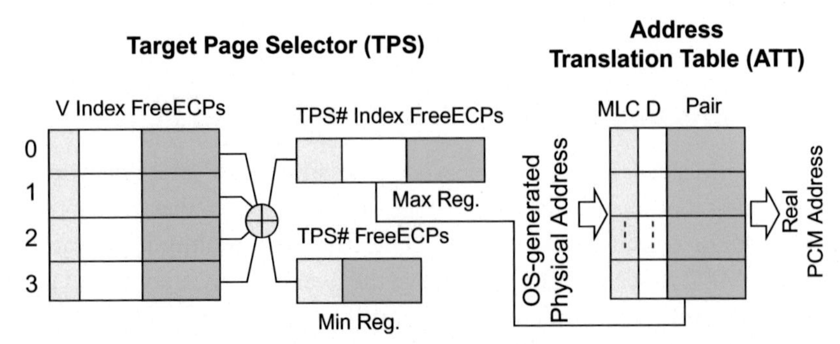

Fig. 2 Structural view of the target page selector (TPS) and address translation table (ATT) units.

in the PCM memory (*Max.Index*). Similarly, a register *Min* with the minimum value of FreeECPs pointers in TPS (*Min.FreeECPs*) and the position of the corresponding entry in TPS (*Min.Entry*) is kept updated. The role of this register is to help replacing the page corresponding to *Min* in TPS with a healthier PCM page.

When the ECP pointers of an unpaired page are updated, the TPS is checked if an update is required. The index of this page is compared in parallel with the indices of valid TPS entries. On a hit, the corresponding entry is updated with the new FreeECPs value. If this value becomes zero, the entry is invalidated, since the corresponding page cannot tolerate any more bit failure and is approaching the end of its lifetime. If the page index does not match an entry in TPS and its number of FreeECP pointers is larger than *Min.FreeECPs*, the TPS entry pointed by *Min.Entry* is updated for this page.

Any change or invalidation in the entries of TPS leads to immediate update of *Min* and *Max* registers. Invalidations take place when the number of FreeECP pointers of a page, which is already in TPS, becomes zero or when the entry pointed by *Max* is used for pairing. In the latter case, FreeECPs register of the entry is also set to zero. This ensures that all invalid entries have the smallest FreeECPs value. Whenever there is at least one invalid entry, *Min* points to it and *Min.Entry* is zero.

TPS indicates the best candidate target page for pairing most of the times. However, it may fail to find the best candidate when all TPS entries are invalidated in the moment that a pairing is requested. In that case, no pairing takes place. We have observed experimentally that a proper size of TPS table for a PCM memory array of K pages is $8 \log K$.

2.2 Pairing algorithm

This algorithm is triggered by FS-OD3P when target page selection algorithm chooses a page for pairing with the faulty page. To this end, the cells of the target page are programmed in 2-bit MLC mode to store data bits of both faulty and target pages. Here, there are two design issues to consider:

- *ECP protection for MLC.* In 2-bit MLC mode, each cell failure causes two bit errors, reducing the efficiency of ECP-6 protection. To mitigate sudden failure of a MLC page, our memory controller allows reallocating the unused ECP pointers of the faulty page for error recovery at target MLC page. Following this approach, it is clear that error recovery capability of the target MLC page (with doubled number of ECP pointers) becomes comparable to a SLC page with ECP-6.

- *Restrictions on pairing*. Although it is the responsibility of target page selection algorithm to choose the target page, it is clear that pairing the faulty page with a target page that has a low number of FreeECP pointers can be counterproductive. Such a situation indicates that the target page will be soon marked faulty and programming it in MLC mode can make it happening even sooner. The target page in this case could continue working well if it was not selected as a target page. A second condition to perform a pairing is that the number of FreeECP pointers of the target page be above a given limit, called *pairing limit*. We have experimentally determined the proper limit is 1 when using ECP-6. To realize a low cost check for this condition, the ECP mechanism can be equipped with a single bit to show if there is at least one free ECP pointer.

One potential drawback to OD3P's pairing mechanism is the possibility of performance loss due to the increased latency of retrieving pages in 2-bit MLC mode. Fig. 3A shows an example of how accesses to any of two PCM pages are performed with conventional stacked mapping. Upon receiving a read to location A (or B), the memory controller issues a read signal to sense the resistance value of each cell. When this value is determined,

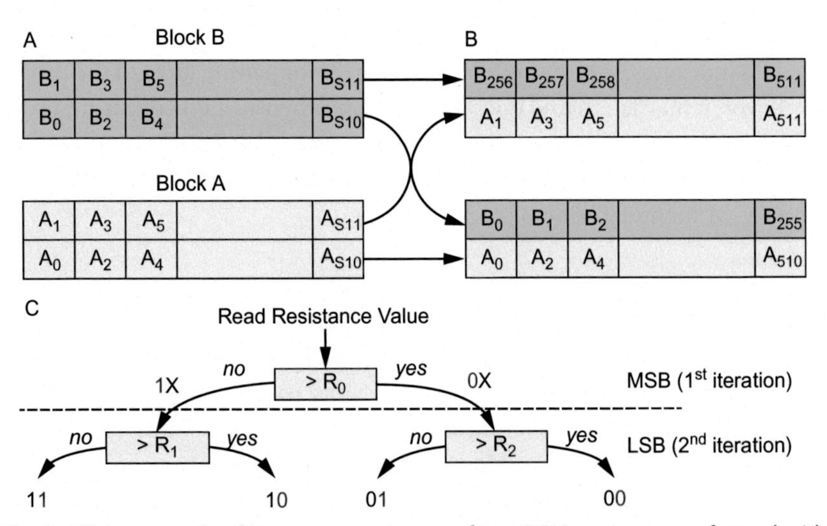

Fig. 3 (A) An example of how accesses to any of two PCM pages are performed with conventional stacked mapping. Upon receiving a read to location A (or B), the memory controller issues a read signal to sense the resistance value of each cell. When this value is determined, the controller uses a binary search to determine the MLC content bit by bit. (B) The circuit compares the resistance value to a reference value roughly fixed at the middle of whole GST resistance range. (C) Interleaved bit mapping for the example of Fig. 3A.

the controller uses a binary search to determine the MLC content bit by bit, as shown in Fig. 3C. This figure shows that at the first step, the circuit compares the resistance value to a reference value roughly fixed at the middle of whole GST resistance range. Depending on the comparison outcome, (1) the most significant bit (MSB) of data is retrieved and (2) the circuit fixes the next resistance reference to extract the least significant bit (LSB). MLC read latency is realized in a sequence of *one sense followed by two comparisons* that takes longer than SLC read comprising of *one sense and one comparison.* One can reduce this latency by comparing the resistance value with all references in parallel (as a flash ADC). However, complexity of a parallel search mechanism is high that limits its practical use for n-bit MLCs.

It is intuitive that we could reduce the latency of one of the paired pages by leveraging the access latency asymmetry of MSB and LSB. This is known as page interleaving and inspired from data arrangement in MLC flash memories [3]. To accomplish this, memory controller arranges data bits of the target page to be interleaved in MLC cells as MSBs. Then, the memory controller can hand reads to the target page by *one sense and one comparison* sequence, resulting in a latency equal to SLC read. Also, data bits of the faulty page are interleaved as LSBs that requires *one sense and two comparisons* to be read. Fig. 3B illustrates interleaved bit mapping for the example of Fig. 3A.

2.3 Address translation

It is required to translate physical memory addresses to real PCM addresses. To facilitate it in FS-OD3P, we introduce the *Address Translation Table* (ATT). Each page in the physical address space is mapped to an entry in ATT in which the pairing information of the page is stored. Each entry consists of a *Mlc* bit which indicates whether cells of the page are in MLC mode (i.e., "0" for SLC mode and "1" for MLC), a *pairing* bit (*P*) which is set when the page is paired, and the address of target page with which it is paired (*Pair*).

Once a faulty page is paired with the target page, *P* is set to "1" and *Pair* tag is filled with the index of target page given by *Max* register in TPS (exactly *Max.Index* field). Also, the *Mlc* bit of the target page is set to "1" referring to its MLC mode for future references.

Notice that if Target Line Selection Algorithm fails to return a page index as target, the faulty page is marked faulty. Also, in this case, both *P* and *Mlc* bits are set to show that the page is faulty and not paired with any target page. Table 1 shows how different bits and tag of an ATT entry can be used to decode the real PCM address of a page.

Table 1 Operations of a faulty page (A) and its target (B) in FS-OD3P coded by fields of ATT.

Mlc	P	Pair	Operation
"0"	"0"	Don't care	Not paired or displaced, stored in SLC
"0"	"1"	Address_B	Displaced to B. Stored in LSBs of 2-bit MLC
"1"	"0"	Don't care	Target. Stored in MSBs of 2-bit MLC
"1"	"1"	Don't care	Not paired, displaced to disk

Since PCM wear-out is a permanent fault, the ATT should be non-volatile to keep pairing information persistent. So, it can be kept in a separate PCM array for fast accesses.

The entries in the table are updated only when a page is paired and they take substantially less wear than the main PCM pages. To mitigate the effect of ATT wear-out, we store three independent replicas of the table. In cases where the entry in the primary copy of the table is faulty, the system tries the second replica and then the third; when all three replicas are not usable, the corresponding physical page is marked as faulty. For a PCM memory with 64-bit address and 4KB page size, three copies consume 198 bits for every 4KB page, i.e., about 0.60% PCM storage overhead.

2.4 Discussion

The main objectives of FS-OD3P are to select the healthiest page in the PCM memory and to provide a flexible mechanism that allows aggressive pairing. The latter case permits if a paired page (in MLC mode) reaches its endurance limit, OD3P may re-pair both memory pages of the faulty page with two new healthy target pages (if remapping and pairing algorithms allow). Once pairings are realized, ATT entries of the two faulty pages as well as their new target pages are updated with new pairing information.

Despite its advantages, the main design drawback with FS-OD3P is the high complexity of TPS. More accurately, it takes a long time for bits of a page to wear-out, but once they do, several stuck at faults occur (due to multi-bit nature of wear-out errors). This results in frequent accesses to TPS table and peripheral circuits. To address this problem, simple alternative schemes for page pairing are discussed next.

3. Fixed pairing algorithm

In contrast to FS-OD3P, the simplest pairing approach is Fixed Pairing (F-OD3P) in which each memory page can be statically paired with one specific target page in the PCM. Obviously, this design reduces the additional hardware complexity of FS-OD3P as well as its searching latency. We have assumed to pair the faulty page with a target page, which is farthest to the faulty page in the PCM memory; this is achieved by complementing the page index of the faulty page.

Such a selection scheme can be justified by the principle of spatial locality of write requests; when a given page is written, write requests to neighboring pages in upcoming cycles are very likely realized. A consequence of this decision is that, given two pages A and B paired by F-OD3P, A can be a target of B when page B fails, and vice versa.

3.1 Pairing algorithm

Just as in FS-OD3P, the actual pairing takes place by programming the target page in 2-bit MLC mode and interleaving data bits of both pages (MSBs hold target page and LSBs hold faulty page). Besides, fixed pairing mechanism relies on two other design issues described in Section 2.2: (1) to enhance reliability of the target page, memory controller uses all ECP pointers of the faulty page to support the target page; (2) pairing algorithm should still check the number of free ECP pointers at the target page to be less than *pairing limit*. If the number of free pointers is below this limit, pairing takes place; otherwise, the faulty page is marked faulty since there is no other choice for pairing in F-OD3P.

3.2 Address translation

Just as in FS-OD3P, the fixed pairing approach uses ATT for physical-to-real address translation. However, the only difference with respect to FS-OD3P is that ATT entries do not contain *Pair* tag field since the possible target page is fixed and inherently known for memory controller. Therefore, ATT has 2 bits (*Mlc* and *P*) for every 4 KB page, resulting in a storage overhead of about 0.006% of PCM storage (assuming we have three replicates of ATT for more reliability). Table 2 shows different encodings of ATT bits for F-OD3P mechanism.

Table 2 Operations of a faulty page and target page in F-OD3P using ATT.

Mlc	P	Operation
"0"	"0"	Not paired, stored in SLC
"0"	"1"	Paired to the target page (stored in LSBs of target MLC line)
"1"	"0"	Target page (stored in MSBs of MLC line)
"1"	"1"	Marked faulty and written on disk

Contrary to FS-OD3P that needs a complex search structure (i.e., TPS), F-OD3P needs not search for a target page and it is known by default. However, it is not desired in most applications since it shows no flexibility in finding a target page for a faulty page, while this flexibility was at its maximum level in FS-OD3P. Such flexibility can greatly help to improve PCM lifetime.

4. Partially-selective pairing algorithm

Based on the above discussion, ideally we would like to have a pairing algorithm with the flexibility of FS-OD3P in choosing most healthy pages for pairing, along with the simplicity of F-OD3P. This motivates the use of a line pairing algorithm with limited selection capability (known as Partially-Selective OD3P or PS-OD3P) obtaining the freedom of FS-OD3P in most cases while still having an acceptable level of complexity (with respect to F-OD3P).

Conceptually, PS-OD3P divides the main memory into several groups with limited pages in each (e.g., 2, 4, 8 or more) with two distinctive features: (1) within a group, we have the maximum freedom to select the most healthy page, like FS-OD3P, but for a limited number of pages. (2) Groups are isolated and no page pairing is allowed for two pages from different groups. (3) Finally, to minimize the impact of write locality as in F-OD3P, pages within a group are selected such that they are far from each other. So, for the group size of G, the target pages for a faulty page are those with different most significant logG bits.

4.1 Address translation

Just as in FS-OD3P, each ATT entry consists of two *Mlc* and *P* bits as well as a *Pair* field. Moreover, operations shown in Table 1 are also valid for PS-OD3P. Note that in the current design, *Pair* field has a size of logG bits (for a group size of G pages), which replace the most significant logG bits of the physical address of the faulty page to achieve the real PCM address.

Based on our experiments in Section 8, groups with 32 and 64 pages provide the best performance-complexity trade-offs with a storage overhead of 0.021% and 0.24%, respectively, for a PCM memory using 4 KB pages.

5. Operation of different OD3P mechanisms: Examples

Fig. 4 shows a simple example scenario of OD3P operation in a small PCM memory of eight pages. Memory line addresses of 4 bits are used for simplicity, with 3 MSBs indicating the page index and 1 LSB to indicate the line within a page (two lines per page). Each page has two PCM lines each

A

Step 1

	Tag	P	MIc	Line 1		Line 0	
0		0	0	0_1	0	0_0	5
1		0	0		4		6
2		0	0		2		4
3		0	0		3		6
4		0	0		1		1
5		0	0	5_1	6	5_0	5
6		0	0		2		6
7		0	0		4		6

Step 2

	Tag	P	MIc	Line 1		Line 0	
0	5	1	0		0		5
1		0	0		4		6
2		0	0		2		4
3		0	0		3		6
4		0	0		1		1
5		0	1	0_1 & 5_1	0	0_0 & 5_0	1
6		0	0		2		6
7		0	0		4		6

Step 3

	Tag	P	MIc	Line 1		Line 0	
0	7	1	0		0		5
1		0	1	5_1&1_1	4	5_0&1_0	6
2		0	0		2		4
3		0	0		3		6
4		0	0		1		1
5	1	1	0		0		1
6		0	0		2		6
7		0	1	0_1&7_1	4	0_0&7_0	6

Fully-Selective OD3P Scheme

B

	P	MIc	Line 1		Line 0	
0	0	0	0_1	0	0_0	5
1	0	0		4		6
2	0	0		2		4
3	0	0		3		6
4	0	0	4_1	2	4_0	2
5	0	0		6		5
6	0	0		2		6
7	0	0		4		6

	P	MIc	Line 1		Line 0	
0	1	0		0		5
1	0	0		4		6
2	0	0		2		4
3	0	0		3		6
4	0	1	0_1 & 4_1	0	0_0 & 4_0	1
5	0	0		6		5
6	0	0		2		6
7	0	0		4		6

	P	MIc	Line 1		Line 0	
0	1	1		0		5
1	0	0		4		6
2	0	0		2		4
3	0	0		3		6
4	1	1		1		1
5	0	0		0		1
6	0	0		2		6
7	0	0		4		6

Fixed OD3P Scheme

C

	Tag	P	MIc	Line 1		Line 0	
0		0	0	0_1	0	0_0	5
1		0	0		4		6
2		0	0	2_1	2	2_0	4
3		0	0		3		6
4		0	0		2		2
5		0	0		6		5
6		0	0		2		2
7		0	0		4		6

	Tag	P	MIc	Line 1		Line 0	
0	2	1	0		0		5
1		0	0		4		6
2		0	1	0_1 & 2_1	0	0_0 & 2_0	2
3		0	0		3		6
4		0	0		2		2
5		0	1		0		1
6		0	0		2		2
7		0	0		4		6

	Tag	P	MIc	Line 1		Line 0	
0	4	1	0		0		5
1		0	0		4		6
2	6	1	0		0		3
3		0	0		3		6
4		0	1	0_1 & 4_1	2	0_0 & 4_0	2
5		0	0		0		1
6		0	1	2_1 & 6_1	2	2_0 & 6_0	2
7		0	0		4		6

Partially-Selective OD3P Scheme with Group Size of 4

Fig. 4 A simple example scenario of OD3P operation in a small PCM memory of eight pages. (A) Fully-selective OD3P scheme, (B) Fixed OD3P scheme, (C) Partially-Selective OD3P scheme with group size of 4.

protected by an ECP-6 scheme (including a tag holding the number of FreeECP pointers). In what follows, we explain what happens in the memory when using different OD3P mechanisms.

FS-OD3P (Fig. 4A): At a given time, a write reference to page 0, where at least one line has FreeECP = 0, leads to detection of an uncorrectable error. After searching for the minimum FreeECP in all pages 1 to 7 and checking if pairing with the selected page (page 5) is allowed ($FreeECPs_{min} > 1$), a pairing of data blocks of page 0 and page 5 occurs. Now, data blocks of page 0 are stored on LSBs of cells in page 5 (in MLC mode); so retrieving pages 5 and 0 can be realized in SLC and MLC read latency, respectively. When an uncorrectable error occurs at page 5 (holding the data blocks of pages 5 and 0), the line pairing algorithm is triggered again to search throughout the PCM memory for two healthy pages far from their lifetime limit ($FreeECPs_{min} > 1$). As a result of the search, pages 0 and 5 are paired with pages 7 and 1, respectively. Then, FS-OD3P tries to sustain the maximum PCM capacity as long as the pairing algorithm knows there is at least one SLC line with (FreeECP > 1).

F-OD3P (Fig. 4B): The first uncorrectable error at page 0 triggers the pairing algorithm to check its fixed target page (i.e., page 4) to have lines with FreeECPs larger than 1. If successful, page 0 is paired to page 4 and stored on the LSBs of the cells in page 4 (in MLC mode). When the first uncorrectable error at page 4 occurs, pages 0 and 4 are marked faulty as they cannot be paired to any other page (as a fundamental limit of F-OD3P). Therefore, F-OD3P can slightly relieve sudden reduction in PCM capacity.

PS-OD3P (Fig. 4C): In partially-selective mechanism, we assume we have two pairing groups each with four pages that are different in two upper bits of page index. Then, when page 0 gets faulty, it may be paired with a page in its group (i.e., pages 2, 4 or 6). This restricts the search domain, but it is large enough to find a proper target page. In this example, page 2 is selected since the minimum FreeECP of its lines is larger than pages 4 and 6, and it also satisfies the pairing condition ($FreeECPs_{min} > 1$). When page 2 (currently holding the data blocks of pages 0 and 2) encounters an error, the pairing algorithm examines pages 4 and 6 for further recovery. If pairing condition is met, pages 0 and 2 are paired to pages 1 and 3, respectively. Notice that if all healthy SLC pages in a group have $FreeECP_{min} = 1$, they cannot be a target page and further pairing is not allowed even if there are some pages in other healthy groups and with $FreeECP_{min} > 1$.

6. Line-level OD3P

Up to now, we assumed each line of a page is equipped with ECP to tolerate six permanent bit errors, and when the first line encounters the 7th bit failure, the line and corresponding page are marked as faulty. The first line failure invalidates a page, while other lines of the page may be far from their lifetime limit. This may cause a sudden drop in memory capacity, in particular when variation of endurance limit of the cells in a page is large.

To relax this problem, we can rely on FREE-p protection mechanism [2] and when a memory line reaches its endurance limit, it can be redirected to a healthy spare line, continuing to normal operation of the memory without invalidating the whole page. In addition, each page has a limit to use a limited number of spare lines beyond which the whole page is marked faulty and the proposed OD3P scheme is invoked. Unfortunately, spare lines in FREE-p are also part of the PCM memory. Then, merging FREE-p at line-level with OD3P at page level may greatly increase the complexity of PCM memory functionalities (read, write, wear-leveling).

We propose the line-level OD3P as an alternative mechanism where a whole page is not marked as faulty at the time a line reaches its endurance limit. Instead, we pair the faulty line with another healthy line in the same page. A page is marked as faulty and is paired with another page when we cannot find any healthy SLC line to host the faulty line. Fig. 5 illustrates an example for line-level pairing where line *i* reaches its endurance limit and ECP-6 cannot protect it anymore. At this point, the controller invokes the line-level OD3P that selects line *j* as the healthiest line (the one with maximum FreeECPs) and stores them (*i* and *j*) in line *j* in 2-bit MLC mode.

Here, note that (1) we use a fully-selective mechanism for pairing at line-level. It is because the number of lines in a page is limited and it is practical to check FreeECPs of all lines to find the most healthy one for pairing, (2) just as page level OD3P, ECP pointers of a faulty paired line can be used for

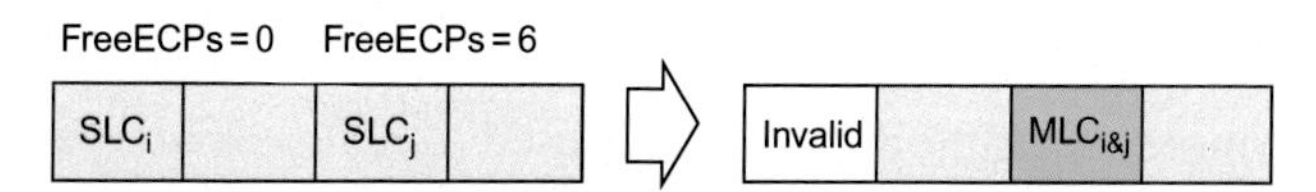

Fig. 5 Example for line-level pairing where line *i* reaches its endurance limit and ECP-6 cannot protect it anymore. Thereafter, invoking the line-level OD3P can select line *j* as the healthiest line (the one with maximum FreeECPs) and store them (*i* and *j*) in line *j* in 2-bit MLC mode.

further protection of the target line, and (3) line-level and page level OD3P mechanisms should not be applied to one page at the same time.

In other words, if lines of a page are paired at some time, the pairing algorithm should not select it as a target page. Fortunately, the proposed design for TPS structure automatically ignores these pages since they have (FreeECP=0). Also, when a page is paired with another one, it cannot use line-level OD3P anymore, since its PCM cells are already used in MLC mode.

7. Simulation environment and scenarios

In this section, we describe the experimental methodology as well as the baseline CMP system used for our evaluation throughout this chapter, unless otherwise noted. We then present the results extracted from simulation experiments in Section 8.

7.1 Infrastructure

We perform micro-architectural simulation of an out-of-order processor model with Ultra SPARCIII ISA using GEMS simulator [4] based on SIMICS toolset [5]. We use CACTI 5.3 [6] for detailed area, power and timing models of cache hierarchy and PCM main memory. We also assume a tiled floor plan of 4-core Ultra SPARC T1 CMP taken from [4] with a TDP of 54 W at 4 GHz and 1.3 V supply voltage.

7.2 System configuration

We model a baseline 4-core tiled CMP. Each tile consists of a SPARC-v9 core (running Solaris 10 OS at 4 GHz), its private L1 instruction/data caches, and one bank of the shared L2 cache with UCA substrate. Table 3 shows parameter settings of the evaluated system.

Coherence among the L1 caches interconnected by a packet-switched network is maintained using a distributed MOESI directory protocol. To speed up simulations, we model an ideal packet switched network with accurate link and router latencies.

Furthermore, deterministic routing is used to reduce router complexity and make energy to have a direct relation to hop counts. The main memory in baseline system is accompanied with a 128 MB shared DRAM cache for improving PCM lifetime and write latency [7]. The baseline has a 4 GB PCM memory with one channel, one rank and sixteen $\times 32$ chips per rank

Table 3 Parameters of the evaluation platform.

Processor	4-Core SPARC-v9, 4GHz, Solaris 10 OS
L1 cache	Split I and D cache; 32 KB private; 4-way; 64B line size; LRU; write-back; 1 port; 2 ns latency
L1 coherency	MOESI directory; 4 × 2 grid packet-switched NoC; XY routing; 3 cycle router; 1 cycle link
L2 cache	UCA shared; 16-way; 64B line size; LRU; write-back; 8 ports; 4 ns latency
DRAM cache	Shared; 128 MB; 4-way; 64B line size; LRU; write-back; 8 ports; 26 ns latency
Off-Chip main memory	4GB MLC; 1 channel; 1 Rank; 16 × 32 chips; 4 banks per chip; 16-entry 64B-size read/write DRAM buffer; 64B row buffer size
Flash SSD	Unlimited size; 25 μs latency

(i.e., a 64 B interface). Each chip is organized into four banks to improve the overall bandwidth.

The memory controller is structured as either for 2-bit MLC evaluation or 3-bit and 4-bit MLCs and is equipped with a 16-entry write queue (WRQ) and a 16-entry read queue (RDQ) for scheduling [8]. In general, the read queue has a higher priority than the write queue. However, when the write queue is 80% full, writes are serviced ahead of reads. In our experiments, we assume the access width of the main memory is equal to the size of a single last level cache line (i.e., 64 B and determines row buffer size) and accesses are sent to the buffers on DRAM cache miss.

7.3 MLC PCM array model

PCM array is similar to the conventional memory structured with DRAM technology. We use CACTI (version 6.5) [6] slightly modified to precisely model the vertically aligned MLC PCM model, the proposed access logic, and peripheral access circuit. As the baseline, we use a 90 nm 2-bit MLC PCM prototype [9] using SET pulse with 250 μA amplitude and 150 ns duration and RESET pulse with 150 μA amplitude and 40 ns duration. To program intermediate states, we assume a total of 32 possible P&V steps with 15 ns current pulse duration and 15 ns switch-off time.

Table 4 Parameters of the 2-bit MLC baseline.

Read	40 μA pulse, 15 ns duration, 15 ns iteration
RESET	250 μA pulse, 150 ns duration
SET	150 μA pulse, 40 ns duration
Partial RESET	0.1 mA pulse with 28 μA stepping, 15 ns pulse, 15 ns Toff, up to 32 P&V steps

These partial SET pulses start from 0.1 mA with 28 μA stepping. The read operation needs 15 ns pulse with 40 μA amplitude for sensing and 15 ns latency for each read iteration [10].

Table 4 gives details of the baseline 2-bit MLC model. Based on the report given in [9,11] this PCM model has $4.8F^2$ cell area. Also, we used the ITRS PID projection model [11] alongside the 4-bit model in [12] to extract values for 3-bit and 4-bit MLCs from 2-bit MLC model parameters.

7.4 Workloads

For workloads, we use both synthetic and real multi-thread and multi-program workloads. Synthetic workloads are used for capacity versus lifetime analysis and are generated with different temporal and spatial write traffics. Also, we use the complete set of parallel workloads provided in PARSEC-2 [13] as multi-thread workloads and SPECCPU2006 benchmarks [14] for multi-program workloads. We classify a benchmark based on the working set size of the programs, read and write intensity of accesses to the main memory (in terms of Read-Per-Kilo-Instructions, and Write-Per-Kilo-Instruction), as well as value locality of its top four most frequent values (MFVs) at 4-bit granularity. Each benchmark is classified by measuring the L2 misses when running alone in the 4-core system of Table 3.

For the multi-program workload selection, we used nine 4-application workloads chosen such that we have a variety of the workloads with different levels of memory intensity. Each application is simulated for 8 billion instructions within its parallel region (i.e., Region of Interest, ROI) and the first 10% of the simulated instructions are assumed as warm-up. Regarding input sets, we use large set for PARSEC-2 applications and sim-large for SPECCPU2006 workloads.

Table 5 characterizes the evaluated workloads based on intensity of their value locality for the baseline system in Table 3.

Table 5 Classification of the evaluated multi-thread (MT) and multi-program (MP) workloads based on intensity of memory access and value locality (VL).

	Program	Memory RPKI	Memory WPKI	Value locality
PARSEC-2	Freqmine	0.704	0.552	62.17%
	X264	4.192	1.832	60.52%
	Dedup	1.281	1.025	72.97%
	Streamcluster	6.791	6.46	82.86%
	Facesim	1.573	2.021	75.88%
	Vips	3.311	7.703	68.42%
	Bodytrack	8.462	0.885	68.45%
	Fluidanimate	5.198	6.125	77.66%
	Ferret	10.018	11.89	75.35%
	Swaptions	1.66	0.317	65.02%
	Raytrace	2.156	0.605	79.81%
	Canneal	12.673	16.257	64.13%
	Blackscholes	2.404	2.173	63.22%
SPEC CPU 2006	MP1: Xalancbmk, omnetpp, bzip2, mcf	0.269	0.211	83.48%
	MP2: Milcm leslie3d, GemsFDTD, lbm	4.106	1.794	79.81%
	MP3: Mcf, xalancbmk, GemsFDTD, lbm	4	3.2	80.77%
	MP4: mcf, GemsFDTD, povray, perlbench	24.395	23.205	67.98%
	MP5: mcf, xalancbmk, perlbench, gcc	0.744	0.956	70.57%
	MP6: GemsFDTD, lbm, povray, namd	3.187	7.413	66.28%
	MP7: gromacs, namd, dealII, povray	4.617	0.483	55.21%
	MP8: perlbench, gcc, dealII, povray	7.942	9.358	61.27%
	MP9: namd, povray, perlbench, gcc	8.094	9.606	58.74%

7.5 Metrics

The metrics used are memory access latency, system performance (Cycles per Instruction, CPI), energy dissipation, and lifetime, for a wide range of device densities including 2-bit, 3-bit, and 4-bit MLCs (2-bit prototypes are now getting popular and 3-bit and 4-bit products are projected to be released in near future).

For energy dissipation, CACTI gives the static power and energy dissipation per each access. So, we can multiply all accesses by the energy of each access, and then divide it by the simulation cycles and get the dynamic power of the memory system. Again, Table 5 illustrates the characterization of the evaluated workloads based on intensity of their value locality for the baseline system in Table 3.

$$\text{Total Power}_{\text{memory unit}} = \text{Leakage Power}_{\text{memory unit}} + \frac{\#\textit{Write Acc.} \times \textit{Energy per Write Acc.} + \#\textit{Read Acc.} \times \textit{Energy per Read Acc.}}{\textit{Simulation Cycles}}. \quad (1)$$

The main endurance metrics used for the evaluated systems are time-to-failure in synthetic analysis and memory lifetime in real workloads. Time-to-failure is defined as the time elapsed between the system startup time to the time PCM capacity is reduced to less than 50% of its maximum capacity (i.e., 2 GB in our 4 GB system). For memory lifetime, we have the same limit for defining system downtime. We assume the number of reliable writes onto a 2-bit MLC PCM cell is limited to 10^6 [11] and by having perfect wear-leveling to simplify the lifetime analysis; so, we have:

$$\text{Lifetime(Year)}_{\text{Memoryunit}} = \frac{\#\text{Maximum reliable write counts}}{\left(\frac{\#\text{Measured writes}}{\text{Simulation cycles}}\right)} \times \frac{365 \times 24 \times 3600}{\text{f (Hz)}} \quad (2)$$

where f is the processor frequency fixed to 2.5×10^9 Hz in our experiments.

8. Experimental results

For the evaluation, we use the system configuration, workloads, and logic model of the PCM memory described in Section 7. In our evaluations, we implemented and compared the following schemes (refer to "Phase-change memory architectures" by M. Asadinia and H. Sarbazi-Azad, for detailed explanation of the comparison methods):

- ECP-6: a system exactly modeling ECP-6 scheme in 128 B line granularity.

- DRM: a system with DRM technique with 2 KB pages utilizing customized parity for byte level correction.
- DRM + ECP-6: a system with DRM for 2 KB pages using ECP-6 correction for 128B lines.
- FS-OD3P: a system with both FS-OD3P at page level (2 KB sizes) and line-level OD3P for 128 B lines. In this design, TPS has 128 entries. We also examine 32, 64, 96, 128, 160, 192, 224, 256 entries in our sensitivity analysis.
- F-OD3P: a system with F-OD3P at page level (2 KB sizes) and line-level OD3P for 128B lines.
- PS-OD3P-n: a system with PS-OD3P at page level (2 KB sizes) and line-level OD3P for 128 B lines. Here, n shows the number of pages in each group. The Partially-Selective OD3P has a default 16 group size. We also examine 2, 4, 8, 16, 32, and 64 group sizes in our sensitivity analysis.

8.1 Analysis under synthetic write traffic

In this section, we compare the endurance potential of OD3P memory systems (i.e., F-OD3P, PS-OD3P-16, FS-OD3P) and baseline systems (i.e., ECP-6, DRM, DRM + ECP-6) by evaluating the PCM capacity and lifetime. Next in Section 9, we analyze different PCM main memory systems under multi-thread and multi-programmed workloads. Evaluating how the PCM capacity changes over time for a large memory system is challenging since large endurance limit of 10^8 per cell requires long simulations.

To make this practical, we ignore accurate full-system simulation and adopt a methodology used in prior related studies [1,15] as a practical alternative. In our evaluation, we assume a PCM memory with 2048 pages, each with 1 KB data cells as well as meta-data information (like ECP and ATT). In this evaluation methodology, both OD3P memories and baseline systems are exactly implemented. Unless noted otherwise, we model variability in cell endurance using the same method in our endurance-analysis example in Section 1 (i.e., a normal distribution function with average of 10^8 and a standard variation of 0.25×10^8).

Fig. 6 compares the lifetime versus capacity provided by the six evaluated systems. With a CoV of 0.25, fixed OD3P (F-OD3P) is the first redirection mechanism (not including ECP) that starts decommissioning pages. This fast reduction in memory capacity has two reasons. First, fixed pairing has a default choice for each page failure. The target of a pairing is probably not a good one. Second, the target page is shape shifting into 2-bit MLC

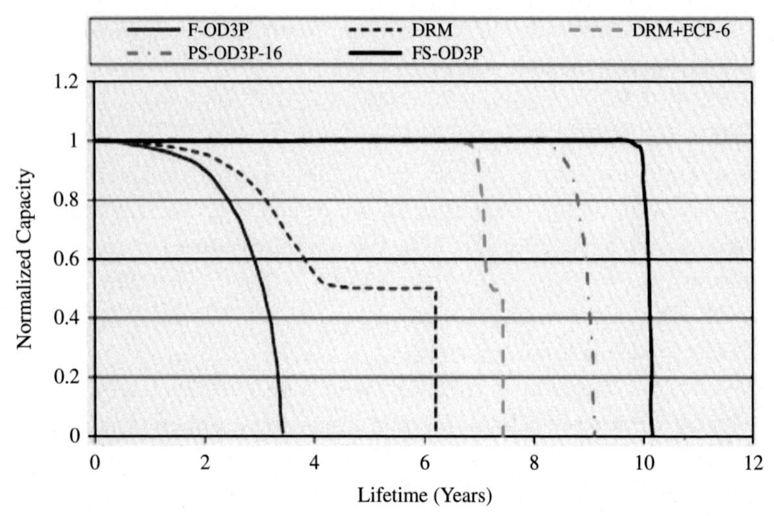

Fig. 6 Comparison of the lifetime versus capacity provided by six evaluated systems with CoV of 0.25.

cells less resilient to hard errors than SLCs. By increasing the domain of target selection for paired line in partially-selective mechanism (PS-OD3P-16), we expect better PCM lifetime and capacity with respect to the fixed pairing mechanism.

This is also observed in Fig. 6 for PS-OD3P of group size 16. The lifetime and capacity results also reveal, in contrast to DRM + ECP-6, PS-OD3P-16 gives more freedom for pairing. This relates to DRM limitation in finding a compatible page for pairing which is less successful.

Finally, FS-OD3P presents the best lifetime versus capacity among all the evaluated systems. However, this lifetime gain is not much larger than that of PS-OD3P-16, implying that under this level of process technology variation, the freedom given by FS-OD3P scheme is beyond the requirements of the imposed write traffic PCM endurance.

We define memory time-to-failure (TTF) as the elapsed time between the start time of memory operation and the time when memory loses more than 50% of its capacity. Fig. 7 compares the TTF for the evaluated systems under CoV of 25%. If we use FS-OD3P up to failure point of all pages, the device retains 50% of its capacity until 76% of the full lifetime of an ideal PCM part, and is not fully exhausted until 82% of the ideal lifetime; furthermore, it presents about 45% improvement over DRM + ECP-6.

To evaluate the impact of high process variation, Fig. 7 gives the lifetime values for 50% capacity in PCM memory with CoV of 35%. This value of

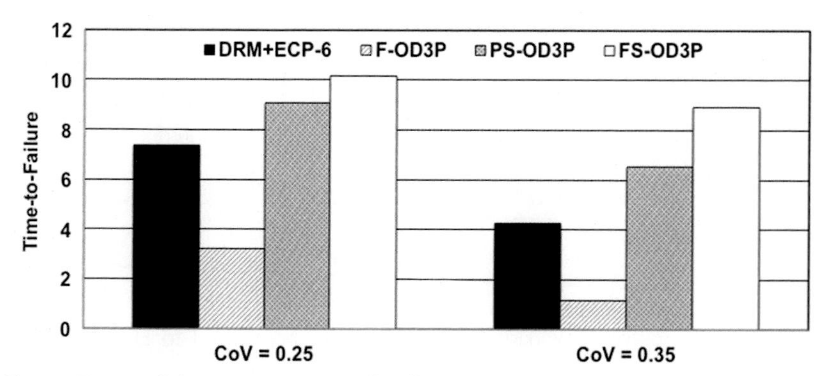

Fig. 7 Time-to-failure comparison of different mechanisms which are based on redirection.

process variation can also be a good representative of imperfect wear-leveling with uneven distribution of write accesses. As expected, FS-OD3P (or PS-OD3P-16 as an alternative) offers better capacity and lifetime gain for this process variation when compared to the other evaluated systems. For instance, FS-OD3P offers a $2.7\times$ longer lifetime than DRM + ECP-6 at 90% capacity and $2\times$ longer at 50% capacity. This increase in lifetime is due to the flexibility provided by the fully-selective pairing scheme compared to more restrictive approaches (like DRM, DRM + ECP-6, F-OD3P, and PS-OD3P-16). More importantly, this CoV of the cell endurance distribution shows the higher endurance enhancement given by the fully-selective mechanism compared to the partially-selective one.

8.2 Analysis under real workloads

We use both multi-thread PARSEC [13] and multi-program SPECCPU2006 [14] workloads for our evaluation under the real workloads.

8.2.1 Performance analysis

Our main performance metric for the evaluated system is speedup, defined as Speedup $= CPI_{ECP}/CPI_{Tech}$ where CPI_{ECP} and CPI_{Tech} are the CPIs of the ECP system (as baseline) and the setting with scheme *Tech*, respectively.

Fig. 8 shows effectiveness of OD3P mechanisms under different workloads. We used a PCM memory system with CoV of 0.25 for cell endurance. We compared OD3Ps (F-OD3P, PS-OD3P-16, and FS-OD3P) with DRM + ECP-6, as DRM + ECP-6 gives the best lifetime and capacity compared to the ECP-based and DRM-based schemes.

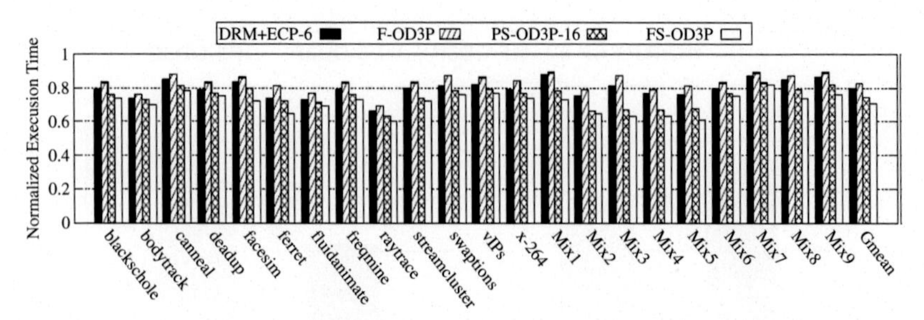

Fig. 8 Performance analysis of OD3P mechanisms compared to DRM + ECP-6 and ECP-6 baselines under different workloads. The values are normalized to ECP-6 baseline system.

From Fig. 8, FS-OD3P's and PS-OD3P-16's lifetime efficiency has a large performance impact. When fully-selective scheme is used, speedup is improved by 14% compared to DRM + ECP-6. We found that PS-OD3P-16 and FS-OD3P have nearly same performance. The reason is a group size of 16 provides enough freedom to maximize the lifetime of the PCM memory. Note that in practice, FS-OD3P is likely to be less efficient than PS-OD3P-16. This is due to the costly search in TPS in terms of latency and energy. Just as in synthetic write traffics, F-OD3P is less efficient in tolerating hard errors when compared to the DRM-based reduction technique.

In Fig. 8, some programs do not utilize the effectiveness of OD3P mechanisms. Here, there are some scenarios: (1) When write traffic is intensive, e.g., in Mix3 and Mix4 applications, the techniques do not necessarily perform better in finding a target page for pairing, (2) When a program has few writes, e.g., in blackscholes and raytrace applications, the capacity reduction due to write operations have little performance impact. (3) When a program has many more reads than writes, e.g., in Mix1 and Mix2 applications, the performance bottleneck shifts from writes to reads such that the pairing mechanism has a small impact (or no impact).

8.2.2 Group size in PS-OD3P

To evaluate the effectiveness of the proposed PS-OD3P scheme with different group sizes, Fig. 9 compares speedup for group sizes ranging from 4 to 64. For each group size, the corresponding bar gives the average values for all 22 real workloads. With a small group size, e.g., 4, there are a large number of page invalidations causing the performance bottleneck being main memory capacity.

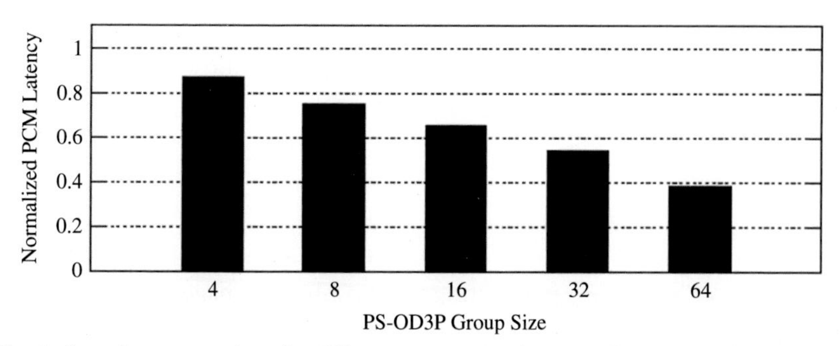

Fig. 9 Speedup comparison for different group size (ranging from 4 to 64).

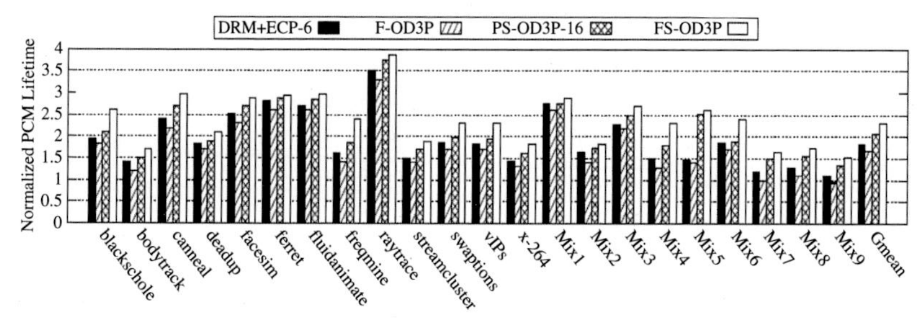

Fig. 10 Endurance analysis of OD3P mechanisms with different configurations and pairing algorithms compared to the baseline architecture. Again, the values are normalized to ECP-6 baseline system.

As group size increases, more freedom in finding the target page is achieved. This improvement in main memory capacity exhibits large performance improvement in cost of increasing the complexity of search mechanism within each group. In our detailed simulations in previous sections, we assumed a group size of 16 since it provides acceptable speedup without severely increasing the search complexity.

8.2.3 Endurance analysis

For lifetime evaluation, we simulate each application up to a time where half of the pages are worn out using different protection mechanisms. At this point, PCM memory reaches its lifetime limit. We measure this elapsed time and depict it against DRM + ECP-6 scheme in Fig. 10.

For failure analysis in synthetic write traffics, we can see that the FS-OD3P mechanism presents the best lifetime among different schemes due to the flexibility it provides in finding the target page. Also, P-OD3P mechanism tries to keep its lifetime high (near to fully-selective one) by increasing the group size to expand the space for finding a good target to pair. Indeed, our proposed method pushes this figure to 15% and 20% improvement in PCM lifetime when using PS-OD3P and FS-OD3P, respectively.

8.2.4 TPS size analysis

TPS plays a great role in our design but we are aware of its impacts. Fig. 11 reports the TPS accuracy in selecting the most healthy page (with respect to the ideal scheme inspecting all memory pages) and TPS area overhead (with respect to total memory area) as a function of TPS size ranging from 32 to 256 entries. For each TPS size, each point on the curve gives the average value for all 22 workloads. As can be seen in the figure, when TPS size increases, more freedom in finding a target page is achieved and the accuracy of TPS is improved. In our simulations, we assume a TPS size of 128 entries since it presents acceptable efficiency in finding the target page (74%) without largely increasing TPS complexity. Note the negligible area overhead of TPS (according to CACTI 6.5).

8.2.5 Performance comparison of OD3P and DRM under different bit failures

To have a better understanding of OD3P architecture and to show when it has better performance than DRM, we performed a set of simulations under different failure conditions. Instead of letting the application's writes and bit-flips cause bit failures, we assumed some bits are failed before running the application and measured the main memory access latency. The bit failure pattern is randomly generated using a three-phase process. First, a random number α ($15 \leq \alpha \leq 35$) is generated to determine the percentage of faulty memory pages in need of pairing. Next, based on the α value, memory pages

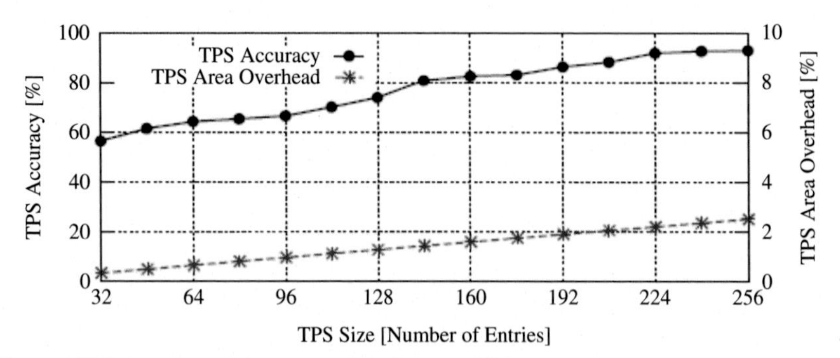

Fig. 11 TPS accuracy and area overhead versus TPS size.

are randomly marked as faulty or healthy. Finally, for a specific number of failed bits (i.e., *FB*), a uniform random function is used to determine whether the position of the failed bits is a faulty page. For a 1 KB page and memory line size of 128B (protected by ECP-6), the minimum number of failed bits (*FB*) is set to 49 $((1\,KB/128\,B) \times 6 + 1)$ to guarantee that a faulty page has at least one failed line.

The maximum number of failed bits is set to 160, as assumed in [16]. For the healthy pages, on the other hand, the bit failure pattern is generated with two features: (1) the number of failed bits per healthy page is set to $FB/2$; and (2) each memory line in a healthy page does not have more than six failed bits (because of 6-bit failure recovery in ECP-6). We selected six programs from our benchmark set with various working set sizes, namely canneal and Mix2 with large working set size; ferret, swaptions, and vips with small working set size; and Mix5 with moderate working set size. As the process is random in nature, we repeated it 10 times and reported the arithmetic average for each benchmark and bit failure value (*FB*).

Fig. 12 shows the access time of DRM normalized to PS-OD3P with group size of 16. We deduced, when the number of failed bits in a faulty page is low, DRM has a better access time to PCM main memory than that of OD3P, while the case is opposite for high bit failures where OD3P is superior to DRM. Besides, the benefits of OD3P over DRM are highlighted when the workload has a large or moderate working set size and we expect OD3P will be much more beneficial when the aggregate size of the applications' working set is large.

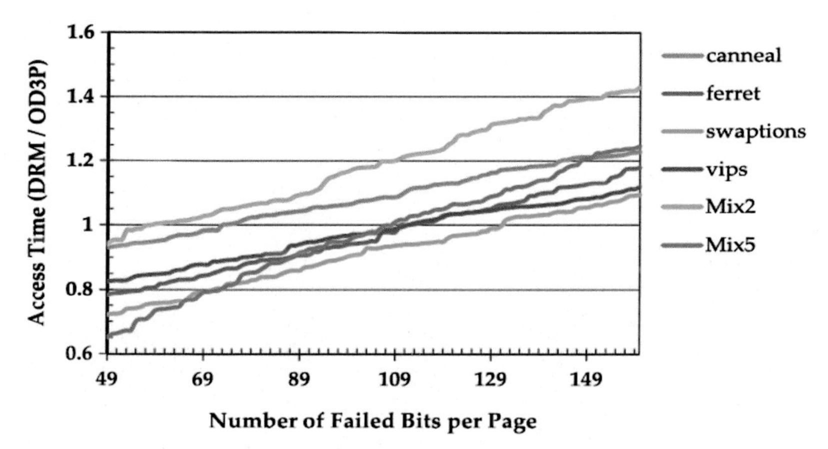

Fig. 12 Access time comparison of OD3P and DRM when number of failed bits varies.

8.3 Comparison to line-level schemes

Now, we compare line-level OD3P with prior techniques proposed to tolerate faults at line (block) level, including SAFER [17], FREE-p [2], LLS [18], and PAYG [19]. Table 6 lists details of the various line-level failure tolerance mechanisms considered here. All these techniques assume a SLC PCM main memory as baseline with the system configuration as shown in Table 3.

Fig. 13 compares the performance and lifetime of different line-level techniques with line-level OD3P under workloads listed in Table 5. The reported values are geometric average values over all workloads. The execution time and lifetime values are normalized to those of OD3P. As can be seen in the table, line-level OD3P provides better execution time and lifetime than all previous schemes. To sum up, one can conclude that line-level OD3P, compared to previous line-level techniques, results in up to 216% average performance improvement while reducing the lifetime by up to

Table 6 Failure tolerance mechanisms at line-level for SLC PCM with their configurations.

Scheme	Approach and configuration
SAFER	Corrects up to 32 errors per block
Free-p	Corrects up to four hard errors per block and remaps block if more errors occurred. Soft-error tolerance is not accounted
LLS	Intra-line salvaging method, such as ECP, which correct initial cell failures. By mapping failed lines to the backup, LLS manages to maintain a contiguous memory space that provides easy integration with wear-leveling
PAYG	Hierarchical ECP with same structure and sizes in the original paper

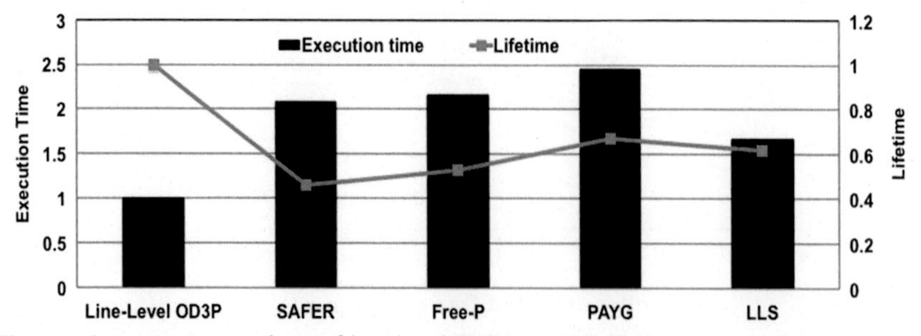

Fig. 13 Comparative analysis of line-level OD3P, SAFER, FREE-P, LLS, and PAYG.

189%. The higher improvements achieved by line-level techniques are due to their ability to use line-level failure tolerance to delay the time a page is marked as faulty within a page, improving the lifetime of the memory. Execution time and lifetime values of each scheme are normalized to those of line-level OD3P and report average values over all workloads.

9. Hardware overhead and extension for N-bit MLC PCM

Almost all applications in all computing segments are requiring larger and larger main memory capacities. SLC PCM memory may become an improper storage scheme and PCM memory systems with high bit densities will be desired. Many previous works considered 2-bit MLC PCM as the technology of choice for main memory systems [10,12,16,20–27].

On the other hand, OD3P assumes a memory system starting with SLC storage level and, when a page/line gets faulty, a target page/line is shape shifted to 2-bit MLC storage level. Read/write circuits must support both SLC and 2-bit MLC accesses, thereby adding a small circuit overhead for OD3P.[a] Here, we show how OD3P can be generalized for an N-bit MLC baseline memory system to improve its lifetime and performance characteristics with no circuit overhead.

We assumed the memory system starts with N-bit storage capacity per cell and when a page/line gets faulty, it is moved on top of an N-bit healthy page. This leads to using $2N$-bit storage cells at OD3P's target pages/lines. In contrast to the initial design of OD3P (with SLC PCM baseline and 2-bit MLC PCM target pages/lines), the generalized N-bit OD3P imposes no implementation cost to access circuits. To better understand this, we show how an N-bit MLC cell can be configured to support $2N$-bit access with negligible implementation overhead. As shown in Fig. 14, an MLC PCM write requires a full-sweep RESET pulse (for programming into RESET state, or initializing the GST for a P&V process), a full-sweep SET pulse (for programming into SET state), and a ProGraMmable pulse (PGM, generating partial SET pulses during P&V) [28]. The PGM unit employs a

[a] In [28], the access logic supporting SLC and 2-bit MLC read/write operations is discussed. They showed that the main contributors to the area of the access circuit are: (1) sense amplifiers and (2) NMOS transistors of the write circuit. The second contributor is much more important in PCM memories since they need pumping large currents into cell arrays. Calculations of the area of the access circuit in 45 nm technology (using HSPICE and CACTI) show that the additional logic to support MLC operations has an overhead of 30.4% compared to SLC's access logic, which is less than 0.5% of the PCM chip area.

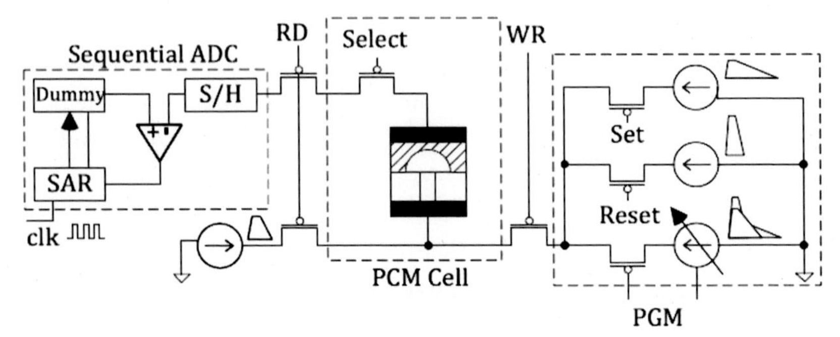

Fig. 14 The PCM cell supporting *N*-bit to 2*N*-bit MLC operations. Dummy block refers to a group of fixed resistive elements used for comparison in a read circuit.

current Digital-Analog Converter (DAC) generating SET pulses with specified width but different amplitudes (64 levels in Fig. 14). In order to increase the density of an *N*-bit MLC PCM to a 2*N*-bit MLC PCM, the PGM control logic determines the amplitude of intermediate SET pulses. For read, on the other hand, the controller doubles the iterations of the sequential ADC according to the target storage level. Therefore, switching from an *N*-bit MLC to 2*N*-bit MLC density does not require additional circuitry, but rather fixing some values in the control logic.

Note that, although *N*-bit MLC PCMs (for $N=3, 4, \ldots$) may be viable in the future, they can badly suffer reliability problems such as short wear-out endurance and high soft error rate due to resistance drift. So, it is really difficult to think about *N*-bit MLC PCMs ($N>4$) in future products.

References

[1] Ipek E, Condit J, Nightingale EB, Burger D, Moscibroda T: Dynamically replicated memory: building reliable systems from nanoscale resistive memories. In ASPLOS; 2010, pp 3–14.
[2] Yoon DH, et al: FREE-p: protecting non-volatile memory against both hard and soft errors. In HPCA; 2011, pp 466–477.
[3] Grupp LM, Davis JD, Swanson S: The Harey Tortoise: managing heterogeneous write performance in SSDs. In ATC, 2013.
[4] Martin MMK, et al: Multifacet's general execution-driven multiprocessor simulator (GEMS) toolset, ACM CAN, vol. 33, 92–99, 2005.
[5] Magnusson P, et al: Simics: a full system simulation platform, *Computer* 35:50–58, 2002.
[6] CACTI: An Integrated Cache and Memory Access Time, Cycle Time, Area, Leakage, and Dynamic Power Model. Ver. 5.3, 2010: Retrieved in June 2010 from http://www.hpl.hp.com/research/cacti/.
[7] Lee BC, et al: Architecting phase change memory as a scalable DRAM alternative. In Proc. Int'l Symp. Computer Architecture (ISCA'09); 2009, pp 2–13.

[8] Qureshi MK, et al: Enhancing lifetime and security of PCM based main memory with start-gap wear leveling. In Proc. IEEE/ACM Int. Symp. Microarchitecture (MICRO'09); 2009, pp 14–23.
[9] Bedeschi F, et al: A bipolar-selected phase change memory featuring multi-level cell storage, *IEEE J Solid State Circuits* 44(1):217–227, 2009.
[10] Joshi M, et al: Mercury: a fast and energy-efficient multi-level cell based phase change memory system. In 7th International Conference on High-Performance Computer Architecture (HPCA); 2011.
[11] ITRS: *International Technology for Semiconductor, Process Integration, Devices, and Structures*, 2011 ed., 2011. Retrieved in April 2012 from: http://www.itrs.net/.
[12] Qureshi MK, Franceschini MM, Lastras-Monta LA: Improving read performance of phase change memories via write cancellation and write pausing. In Proc. IEEE Symp. High Performance Computer Architecture (HPCA'10); 2010, pp 1–11.
[13] Bienia C, Li K: PARSEC 2.0: a new benchmark suite for chip-multiprocessors. In Proc. Workshop on Modeling, Benchmarking and Simulation (MoBS); June 2009.
[14] Spradling CD: SPEC CPU2006 benchmark tools, *ACM SIGARCH Comput Archit News* 35(1):130–134, 2007.
[15] Schechter SE, et al: Use ECP, not ECC, for hard failures in resistive memories. In Proc. Int. Symp. Computer Architecture (ISCA); 2010, pp 141–152.
[16] Azevedo R, Davis JD, Strauss K, Gopalan P, Manasse M, Yekhanin S: Zombie memory: extending memory lifetime by reviving dead blocks. In Proc. Int. Symp. Computer Architecture (ISCA); 2013, pp 452–463.
[17] Seong NH, et al: SAFER: stuck-at-fault error recovery for memories. In *MICRO*; 2010, pp 115–124.
[18] Jiang L, Du Y, Zhang Y, Childers BR, Yang J: LLS: cooperative integration of wear-leveling and salvaging for PCM main memory. In Proc. Int. Conf. Dep. Sys. and Net. (DSN); 2011, pp 221–232.
[19] Qureshi MK: Pay-as-you-go: low-overhead hard-error correction for phase change memories. In Proc. IEEE/ACM Int. Symp. Microarchitecture (MICRO); 2011, pp 318–328.
[20] Nair PJ, Chou C, Rajendran B, Qureshi MK: Reducing read latency of phase change memory via early read and Turbo Read. In Proc. Int. Symp. High Performance Computer Architecture (HPCA); 2015, pp 309–319.
[21] Jiang L, et al: Improving write operations in MLC phase change memory. In *Proc. IEEE Symp. High Performance Computer Architecture* (HPCA); February 2012, pp 201–210.
[22] Jalili M, Sarbazi-Azad H: Express read in MLC phase change memories, *ACM Trans Des Autom Electron Syst* 23(3):1–33, 2018.
[23] Hoseinzadeh M, Arjomand M, Sarbazi-Azad H: SPCM: the striped phase change memory, *ACM Trans Archit Code Optim* 12(4):1–25, 2016.
[24] Hoseinzadeh M, Arjomand M, Sarbazi-Azad H: Reducing access latency of MLC PCMs through line striping. In International Symposium on Computer Architecture (ISCA); 2014, pp 277–288.
[25] Qureshi MK, et al: Morphable memory system: a robust architecture for exploiting multi-level phase change memories. In Proc. Int. Symp. Computer Architecture (ISCA); June 2010, pp 153–162.
[26] Zhang W, Li T: Helmet: a resistance drift resilient architecture for multi-level cell phase change memory system. In Proc. IEEE/IFIP Int. Conf. Dependable Systems and Networks (DSN); 2011, pp 197–208.
[27] Xu W, Zhang T: Using time-aware memory sensing to address resistance drift issue in multi-level phase change memory. In Proc. IEEE Symp. Quality Electronic Design (ISQED' 10); 2010, pp 356–361.

[28] Dong X, Xie Y: AdaMS: adaptiveMLC/SLC phase-change memory design for file storage. In Proc. Asia and South Pacific Design Automation Conference (ASP-DAC); 2011, pp 31–36.

Further reading

[29] Asadinia M, Arjomand M, Sarbazi-Azad H: Prolonging lifetime of PCM-based main memories through on demand page pairing, *ACM Trans Des Autom Electron Syst* 20:1–24, 2015.

[30] Asadinia M, Arjomand M, Sarbazi-Azad H: OD3P: on-demand page paired PCM. In Proc. Design Automation Conference (DAC); 2014, pp 1–6.

CHAPTER FIVE

Handling hard errors in PCMs by using intra-line level schemes

Marjan Asadinia[a], Hamid Sarbazi-Azad[b]
[a]University of Arkansas, Fayetteville, AR, United States
[b]Sharif University of Technology and Institute for Research in Fundamental Sciences (IPM), Tehran, Iran

Contents

Advances in Computers, Volume 118
ISSN 0065-2458
https://doi.org/10.1016/bs.adcom.2019.10.003

Abstract

In this chapter, we first introduce one shifting mechanisms in order to further prolonging the lifetime of a phase change memory (PCM) device, reducing the write rate to PCM cells, and handling cell failures when hard faults occur. In this line, Byte-level Shifting Scheme (BLESS) is addressed and reduces write pressure over hot cells of blocks. Additionally, we illustrate that using the MLC capability of PCM and manipulating the data block to recover faulty cells can also be used for error recovery purpose. Next, we propose another intra-line level pairing scheme (ILP). This novel recovery mechanism can statically partition a data block into a small number of groups and efficiently benefits from the advantages of MLC capability and enables word-pairing mechanism within a line.

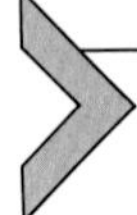

1. BLESS: A simple and efficient scheme for prolonging PCM lifetime

To further prolong the lifetime of a phase change memory (PCM) device, our experiments confirm that during write operations, an extensive non-uniformity in bit flips is exhibited. To reduce this non-uniformity, we present byte-level shifting scheme (BLESS) that reduces write pressure over hot cells of blocks. Also, by utilizing the MLC capability of PCM and manipulating the data block to recover faulty cells, this shifting mechanism can be used for error recovery purpose [1]. In other words, none of the previous works consider the existing non-uniformity of cells activity within a block and they presume a uniform bit flips distribution among all cells.

Fig. 1 illustrates this non-uniformity for one of the multi-threaded application (Facesim). Therefore, it is necessary to propose a method that can reduce this non-uniformity to improve lifetime. In this section, we first introduce our proposed scheme named BLESS. To enhance the PCM lifetime, we must handle two problems: non-uniformity in bit flips and hard errors occurrence. We propose a byte-level shift-based mechanism for distributing bit flips in a block (or line) and reusing healthy cells of a partially faulty block in MLC mode.

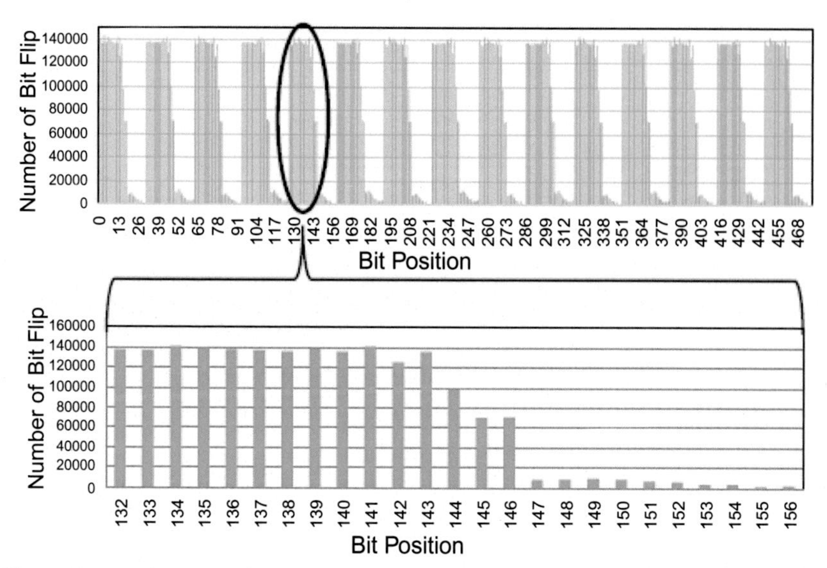

Fig. 1 Non-uniformity in bit flips distribution of write operations for "Facesim."

2. Improving the bit flips uniformity

We observe that there exists an extensive non-uniformity in bit flips during write operations. To improve and extend the lifetime, a good solution is needed in order to somehow reduce this non-uniformity. To this end, we use a byte-level shifting mechanism for each block. In other words, using a 6-bit counter (for 64B blocks), upon each write, we rotationally shift the content of a block by one byte. Doing so provides more uniformity of bit flips across memory blocks.

This is a simple yet effective way with a very low storage overhead. Fig. 2 shows our shifting mechanism. As can be seen, on each write (top to bottom) the content of the block is rotated by one byte. It should be considered that using the shift at lower granularity can also be adopted, but its storage overhead is not tolerable. A sensitivity analysis on the size of shifted units is done in the following sections.

3. Tolerating the hard errors

When hard errors occur, the 6-bit counter used for uniforming bit flips is used for implementing a pairing mechanism that can recover the data

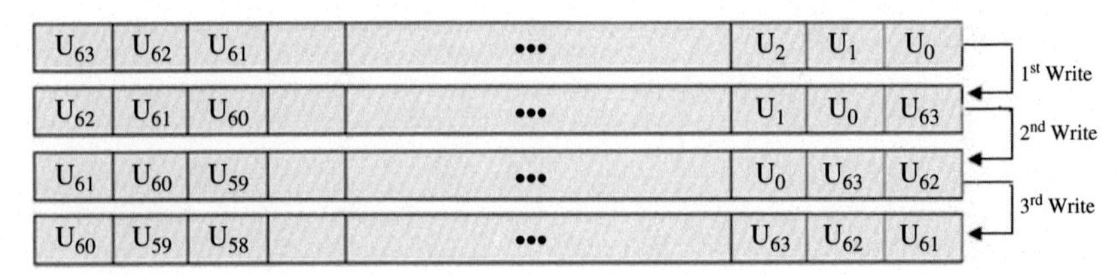

Fig. 2 Shift mechanism: upon each write, the block content is shifted by one unit.

stored in a partially faulty block. To this end, the original block is combined with its shifted form and stored together in MLC mode; the original data block is recovered using the stored data and its shifted form.

3.1 Problem formulation

Let an n-bit binary vector $V = \nu_{n-1}\nu_{n-2} \ldots \nu_2\nu_1\,\nu_0$ define the health status of an n-unit memory block (in our experiments, a unit is one byte (8 bits), unless otherwise mentioned), where $\nu_i = 0$ if the corresponding i-th unit of the block is healthy and $\nu_i = 1$ when it is faulty.

Definition 1. Two vectors U and V are compatible if $U \odot V = u_{n-1}\,\nu_{n-1} + u_{n-1}\,\nu_{n-1} + u_{n-2}\,\nu_{n-2} + \ldots + u_1\,\nu_1 + u_0\,\nu_0 = 0$. Otherwise, they are incompatible.

It is clear that:

(1). Any vector V is compatible with null vector $Z = 00\ldots0$;

(2). Any vector V is compatible with its health-complement vector $\overline{V} = \bar{\upsilon}_{n-1}\bar{\upsilon}_{n-2}\ldots\bar{\upsilon}_1\bar{\upsilon}_0$; and

(3). Any non-null vector V ($V \neq Z$) is incompatible with the completely faulty vector $I = 11\ldots\,1$.

Definition 2. The i-position rotated vector $V^{\rightarrow i}$ is defined as $V^{\rightarrow i} = \nu_{i-1}\nu_{i-2}\ldots\nu_0\nu_{n-1}\nu_{n-2}\ldots\nu_i$.

Definition 3. Vector V is called i-self-compatible if V is compatible with $V^{\rightarrow i}$.

Definition 4. Vector V is called self-compatible if there exists an i, $0 \leq i \leq n-1$, for which V is i-self-compatible.

Corollary 1. *If Hamming Weigh (V)* $> n/2$*, then V cannot be self-compatible.*

Let F_j be the set of all vectors with Hamming weight j and $C_j \subset F_j$ be the set of such vectors that are self-compatible. Then, the probability that a vector with Hamming weight j is self-compatible is given by:

$$P_j = |C_j|/|F_j| = |C_j|/\binom{n}{j} \tag{1}$$

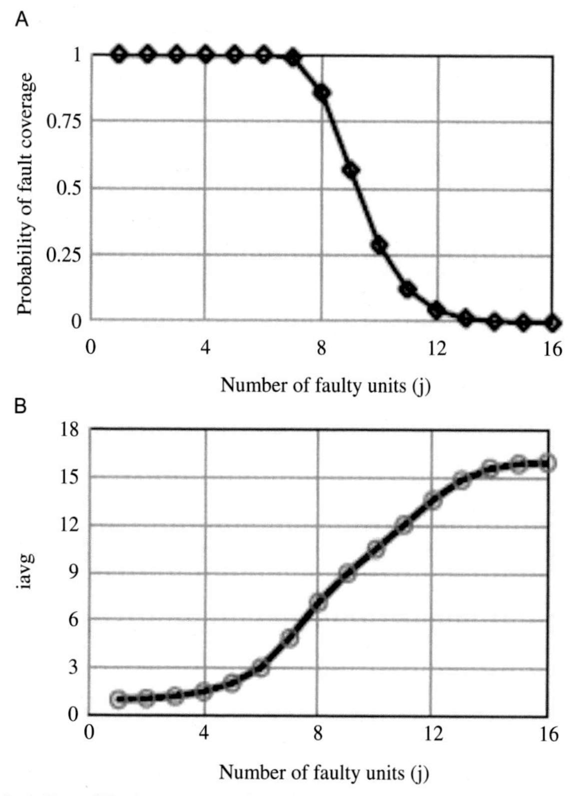

Fig. 3 (A) Probability of fault coverage (Eq. 1). (B) Average number of shifts to cover the fault(s) (Eq. 2).

This means, assuming a uniform distribution of faulty cells in the block, when a block has j faulty units, the probability that the block can be fully recovered is P_j. Fig. 3A shows the probability of fault coverage, P_j, as a function of the number of faulty units in a 16-unit block. As can be seen up to eight faulty segments are recovered in more than 80% of times.

Let $i_{min}(V)$ show the minimum value of i, $0 \leq i \leq n-1$, for which V is i-self-compatible. Then, the average number of shifts required to recover a block with j faulty units is calculated by:

$$i_{avg} = \frac{\sum_{V \in Fj} imin(V)}{|Fj|} \tag{2}$$

Fig. 3B shows the average number of required shifts, i_{avg}, to recover a faulty block as a function of the number of faulty units in the block. Fig. 3 indicates that most faults can be covered by up to eight shifts.

More precisely, when hard errors occur, we distinguish the faulty bytes from the healthy bytes by generating the byte-level fault-mask (same as DRM). Thereafter, we shift the fault-mask until it and its shifted form do not have any "1" at the same position. By a tolerable amount of shift, we can very probably find such a state (a state that fault-mask and its shifted form do not have any "1" in the same position). If such a state cannot be found (due to the pattern of fault-mask), the whole block is decommissioned from the main memory and marked as permanently faulty and will not be used by the memory manager in the future.

If such a state is found, then we combine the original form of the data block and its shifted form (according to the shift count of the fault-mask). Afterward, these two data blocks are stored in one memory block in MLC mode.

To clarify these operations, in what follows, we elaborate our proposal by some example. Fig. 4 illustrates an example of BLESS operation with a 3-unit block, each of 4 bits. A counter keeps the number of shifts (in this example, a 2-bit counter suffices). Now, assume that a write to this block arrives to the memory controller and by using the detection mechanism (i.e., using read-after-write mechanism like many previous works), the failed unit is detected (here, unit 2) and the fault-mask is formed as "100." Now, we shift the fault-mask once and get the new fault-mask "010." In the next step, we compare the original fault-mask "100" to its shifted form "010" and see that they do not have any "1" in the same position. So, they are compatible and the desired state is found by one shift. Shifting the content of the block by one unit, we can reach to the new content, which is the shifted form of the original content. Now, both original data block and shifted data block are combined in MLC mode. Adopting such a scheme ensures us that all bits of the block are stored in healthy cells. To retrieve the content of the block, if we distinguish that the block is programmed in SLC mode

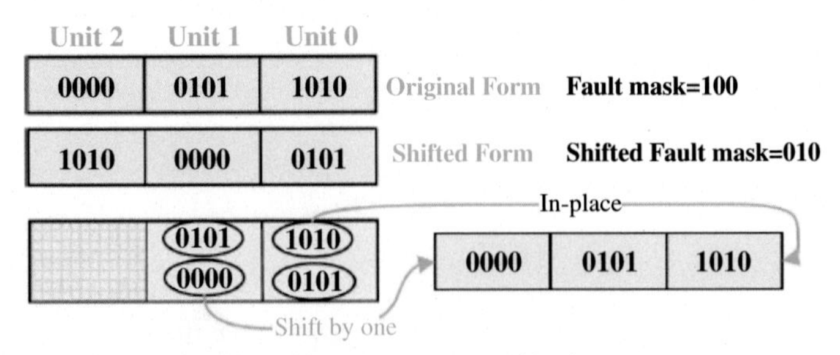

Fig. 4 An example of BLESS operation on a 3-unit block.

(there is a 1-bit indicator associated to each block to show if the block is in SLC/MLC mode), it means that the block is healthy and we can read the block simply in one SLC read cycle. Otherwise, first, we read the block in MLC mode and then prune the data stored in faulty cells and using the stored shift count we can compile the original block content, as shown in Fig. 4. Briefly, the read and write operations are summarized as follows.

4. Write operation in BLESS

When a write request arrives to the memory controller, upon a hard error occurrence, our scheme first determines the byte-level fault-mask of the block. We then shift the fault-mask (a maximum of eight shifts work well) in order to find the state where the fault-mask and its shifted form are compatible. If such a shifted fault-mask is found, it means that we can store the block in its place and the storage type of block is converted to MLC. This way, we shift the content of the block in the extent of shift count and then combine it by the original data block. Afterward, the data is sent to the memory write circuit in order to finish the write operation in MLC mode. When the fault-mask is not self-compatible (with at most eight shifts), then the block is marked as "permanently faulty" and excluded from the memory space to be managed by the memory manager.

5. Read operation in BLESS

If a read request is issued for the memory controller, first, we determine whether the block is stored in MLC/SLC mode. If it is in SLC mode, it means that the block was healthy and one normal SLC read iteration is realized and the data is delivered to the requester. On the contrary, if the block has been stored in MLC mode, it shows that the previous write to the block has been handled by BLESS. Therefore, we set the read circuit to read the content of the block in MLC mode. Then, based on its fault-mask, we obtain the correct content of the block. To do so, as we explained by an example, we ignore the bits that are stored in the faulty positions and try to retrieve the bits from healthy cells. More clearly, the MSBs of each healthy cell are in their right positions, but the LSBs must be shifted by the number given by the shift count to locate them at their right place. After obtaining the block content, the data block is passed to the memory controller to deliver it to read requester.

6. Meta-data information

For each line of 64B, we use one SLC/MLC mode bit. The SLC/MLC bit takes "0" (SLC mode) when the block is healthy and "1" (MLC mode) when pairing is needed. We use a 6-bit counter (for 64B blocks) to implement our shift-based bit flips uniformation in SLC mode; the counter is also used in MLC mode to indicate the number of shifts applied to the block content for pairing purpose. Additionally, to enable our pairing mechanism, as in DRM, for each block, one bit per byte is used to indicate the parity of the byte. Using such a bit (as used in DRM), one-bit faults can be detected (and recovered by checking all 8 bits of the byte). If more than one fault occurs in a byte, then the byte is faulty. Using all such bits for a block forms the fault-mask of the block. The fault-mask is used for finding a proper pair for the faulty block, as explained above. Briefly, we need $1+6+64=71$ bits per block, resulting in $71/512=13.8\%$ storage overhead.

7. Evaluation setting

In this section, we describe our simulation methodology and present evaluation results of the proposed method compared to other known hard error recovery schemes in PCM. We use microarchitectural-level simulation of an out-of-order processor with ALPHA ISA using gem5 [2]. We model the memory controller which gives higher priority to reads and when there is no read, a write is scheduled. The entire memory hierarchy is modeled in the simulator. We model a 4-core 2GHz CMP system with three levels of caches and a PCM main memory. The 8GB PCM main memory is divided into 16 banks. Our system configuration is shown in Table 1. To alleviate the pressure on the PCM main memory bandwidth, the baseline has a 16MB shared write-back DRAM cache. We use the multi-threaded applications from PARSEC in our evaluations [3].

8. Methodology

In this part, statistical simulation approach is performed to evaluate our proposed architecture compared to state-of-the-art schemes over long periods of time. We use a statistical simulator that exploits a methodology utilized in [4] with some modifications to cover BLESS demands.

This simulator consists of 1K memory pages; each page is divided into 64 physical lines and each line consists of 64 bytes. There are some

Table 1 System configuration.

Processor	4-Core ALPHA 21264, 2.0 GHz
L1 Cache	Split I and D cache; 32 KB private; 4-way; 64B line size; LRU; write-back; 1 port; 2 ns latency
L1 Coherency	MOESI directory; 4 × 2 grid packet switched NoC; XY routing; 3 cycle router; 1 cycle link
L2 Cache	4 MB, UCA shared; 16-way; 64B line size; LRU; write-back; 8 ports; 4 ns latency
DRAM Cache	16 MB; 4-way; 64B line size; LRU; write-back; 8 ports; 26 ns latency
Off-Chip Main Memory	8 GB; 16 banks; 64B; open page, SLC: read latency 80 ns (6 ns tPRE + 69 ns tSENSE + 5 ns tBUS); write latency 250 ns
Flash SSD	Unlimited size; 25 μs latency

assumptions for all systems considered in our analysis. First, writing to a page may change only one of its blocks. Second, for each page, we issue writes to the memory system line-by-line to achieve prefect wear-leveling. It is noticeable that all writes to the pages are sequential. More clearly, we issue write#0 to block#0 for all pages, then write#1 to block#1, and so on. For life-time evaluation, we measure the number of writes per page in different approaches. For each cell, we consider a normal distribution with a mean value of 10^8 writes and 0.25 coefficient of variance [4].

For error detection, like most previous works, we use read-after-write scheme instead of using ECC. Since our proposal is a content-dependent scheme, we should consider the fact that for each issued write, we cannot produce random contents. Therefore, for different applications, we use 50 million cache line samples taken from full-system simulations in gem5. In this way, we warmed up the system for 500 million instructions from the beginning of the ROI and then 50 million blocks in the write time are stored separately. Some applications (e.g., Black) are not memory-intensive and do not generate 50 million samples during ROI; so, we use all generated blocks during ROI for these applications.

9. Evaluated architectures

We implemented and evaluated systems with a main memory that uses the following error correction schemes:

- ECP-6: a system that has the ability of correcting a maximum of six hard errors using six pointers.

- SAFER-32: a system using partitioning and inversion technique that can correct up to 32 errors per line.
- Aegis-23 × 23: a recovery solution scheme that uses systematic partitioning by exploiting fewer groups to recover more faults.

  We also consider two other known page-level error correction schemes including:
- DRM: a system that uses parity for error detection and line pairing for byte-level correction.
- PAYG: a system including hierarchical ECP with similar structure and sizes in ECP-6.

10. Evaluation metrics

Our main endurance metric for the evaluated systems is lifetime. We also evaluate the average recoverable errors. Additionally, for taking the reduction in bit flips variation into account, before and after employing BLESS, we use the introduced parameter in [5], called *IntraV*, given as:

$$IntraV = \frac{1}{BFavr.N} \times \sum_{i=1}^{N} \sqrt{\frac{\sum_{j=1}^{512} \left(BFij - \sum_{j=1}^{512} wij/512\right)^2}{511}} \quad (3)$$

where BF_{ij} is the write count of cell j in block i, and BF_{aver} is the average bit flips count and N is the total number of blocks.

11. Evaluation results

In this section, we present our evaluation results of BLESS compared to other hard error recovery schemes.

11.1 Analysis under real workloads

For the evaluation, we use PARSEC multi-threaded workloads [3]. We then compare our proposal in terms of lifetime, number of required shifts, number of recovered errors per page, performance, IntraV and impact of unit size.

11.1.1 Lifetime

As can be seen in Section 5, BLESS outperforms the considered state-of-the-art schemes in terms of lifetime. Assuming the number of writes which reduces the memory capacity to 50% of its initial capacity as a lifetime indicator, BLESS improves it by about 25%, 21% and 15% compared to ECP-6, SAFER-32 and Aegis-23 × 23, respectively (Fig. 5).

11.1.2 Number of required shifts

Fig. 6 reports the distribution of the recovered faulty lines for different number of shifts (the cumulative percentage of recovered lines with 9, 10… 63 shifts is negligible and not shown). As can be seen in the figure, more than 60% of recovered faulty lines are recovered by one shift and about 2% of them are recovered by eight shifts. So, to speed up the process of checking for self-compatibility of a faulty line, we may limit it to up to eight shifts.

11.1.3 Number of recovered errors per page

Table 2 shows the average number of recovered errors per page for different memory systems. Since our system is content-dependent, its average error recovery rate per page depends on the application. As can be seen, BLESS can correct 287 errors in a page which is 91%, 68%, and 7% higher than ECP-6, SAFER-32 and Aegis-23 × 23, respectively.

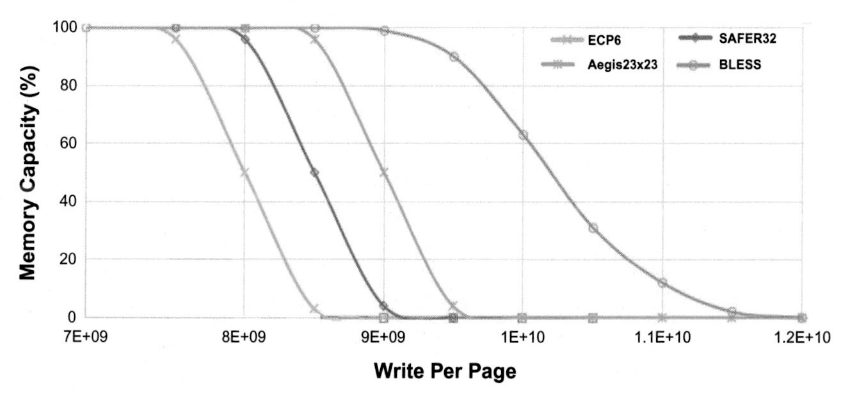

Fig. 5 Lifetime of the BLESS and other techniques.

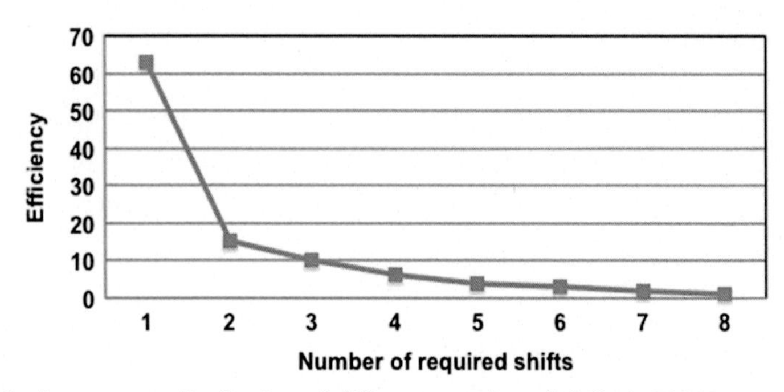

Fig. 6 Fault coverage distribution of different number of shifts in BLESS.

Table 2 Average number of recovered errors per page for different schemes.

Scheme	Average number of recoverable error
ECP-6	150
SAFER-32	170
Aegis 23 × 23	268
Proposed method	287

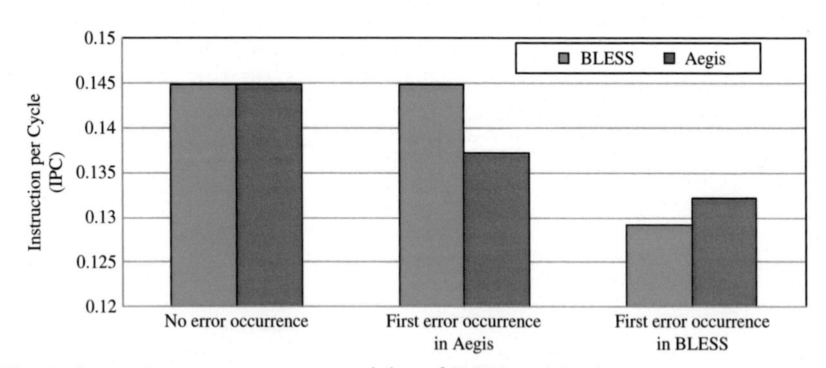

Fig. 7 Comparing error recovery capability of BLESS and Aegis.

11.1.4 Performance

Fig. 7 shows the performance of BLESS compared to Aegis. We consider three regions for our evaluation:

(1). When no error has occurred and both systems are working in an error-free environment. As can be seen in the figure (first bars), both schemes work similar when no error has occurred.

(2). When the first hard error occurs in Aegis, its performance is diminished, while BLESS is still working in its normal way. From this point

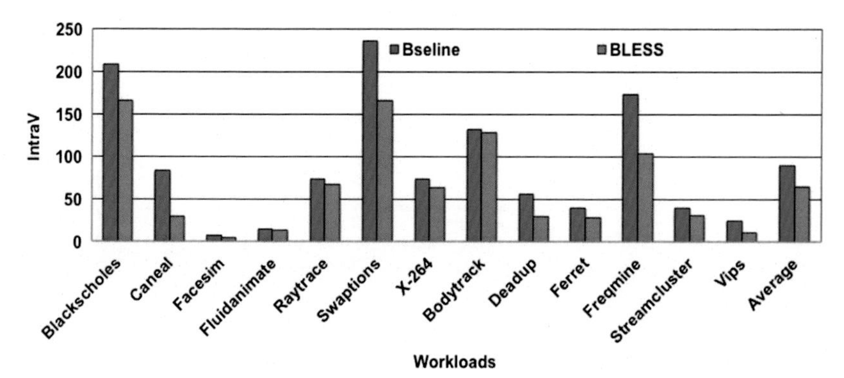

Fig. 8 IntraV of the PARSEC workloads before and after using BLESS.

until the occurrence of the first hard error in BLESS, our scheme works better than Aegis. This is mainly because of the complexity of Aegis operation.

(3). When the first error occurs in BLESS, Aegis is favored and provides a better performance. This is mainly due to long MLC read/write operations in BLESS.

11.1.5 IntraV

Fig. 8 shows the amount of reduction in bit flips for different workloads. As can be seen, our scheme can reduce the variation of bit flips in each block by 27%, on average.

11.1.6 Impact of unit size

The unit size in BLESS plays an important role in determining the storage overhead and IntraV value. Larger units mean less number of units and shorter shift counters and thus less meta-data overhead. However, larger units also reduce IntraV parameter, which is not desirable.

Fig. 9 reports the impact of unit size on storage overhead and IntraV. In this figure, the horizontal axis shows the unit size and left vertical axis shows the meta-data storage overhead per line. More clearly, for each line of 64B, we use one SLC/MLC bit and a 6-bit shift counter to implement bit flips uniformation in SLC mode and to indicate the number of shifts applied to the block content in MLC mode. The right vertical axis shows the IntraV value as a function of unit size. As it is shown in the figure, for 8-bit unit size, we have the best storage overhead and IntraV.

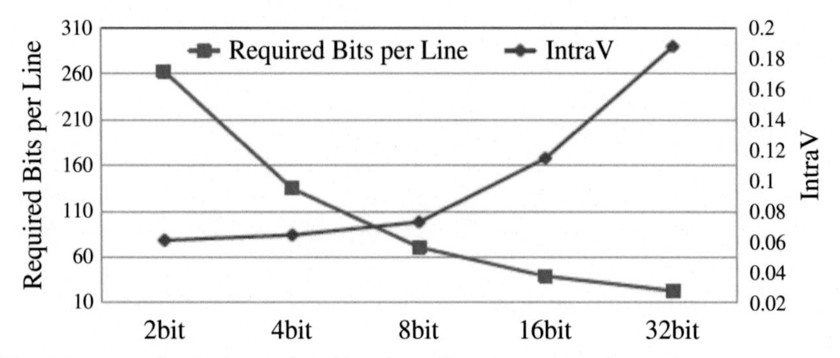

Fig. 9 Impact of unit size on IntraV and per line storage overhead.

12. Comparison to page-level schemes

Here we compare BLESS with DRM [6] and PAYG [7], two known page-level schemes. SLC PCM main memory is considered as a baseline with system configuration shown in Table 1 under multi-threaded workloads. Compared to DRM and PAYG, the average lifetime improvement achieved by BLESS is, respectively, 17% and 14%. Note that BLESS can be coupled with any page-level scheme to further improve the lifetime.

13. Intra-line level pairing (ILP)

As weak cells fail and failed lines in large quantity is present, few techniques are being proposed which can help in increasing the duration of a PCM device when failed lines are remapped to spares and PCM device is salvaged. Others deal with failures through inter-line pairing. The observations we made prove that a line would have much healthy cells even when a line is detected as faulty through the previously presented error recovery mechanisms such as DRM, ECP, FREE-p, and SAFER. It means that when a line is faulty, it can still have many cells that are healthy. This problem could be overcome through ILP or Intra Line Level Pairing [8]. ILP is a method that can solve the fast failure problem of lines by pairing faulty parts of a line onto other healthy parts of the same line. The part that is targeted could be programmed through MLC or Multi Level Cell mode so as to store the data of both parts that are faulty as well as healthy. Therefore, healthy cells could lead to increase in the duration of memory lifetime and lower degradation of capacity [8].

14. ILP structure

As we have suggested, in the case of detection of errors, our pairing scheme compared to other techniques uses read after-write to check the write is done successfully. It is notable that performing read after-write would come at cost of lower system performance due to the fact that read is faster than write. PCM writes also cost high due to latency and energy consumption in contrast to the read.

For error correction, ECP-6 is used at each memory line and capable to correct up to 6 faulty bits per line. Indeed, ECP pointer can save the position of the failed bit and correct its value to then save it in its associated storage bit. When all corresponding pointers of a line are positioned, any error detection lead to declare the entire line is faulty. Therefore, memory controller calls our recovery mechanism.

As an alternative recovery mechanism, intra-line level mechanism (ILP) is proposed where a whole line is not marked as faulty at the time ECP cannot tolerate more faults. To reach this goal, the line is first split into several parts that are paired using faulty parts of the line with some good and healthy parts of it.

When no healthy part is found in order to be paired with a faulty part, the line is pointed as faulty; it is possible to pair this line with other healthy line in the same page in the future.

Fig. 10 describes the intra-line level pairing procedure. A line would reach the limit of endurance where ECP-6 fails. However, here, the controller invokes the pairing mechanism.

Line gets divided into "n" parts where "i" denoted a number of faulty bits where ECP failed to tolerate them. In such circumstances, the part which is healthy should be available to pair with "i."

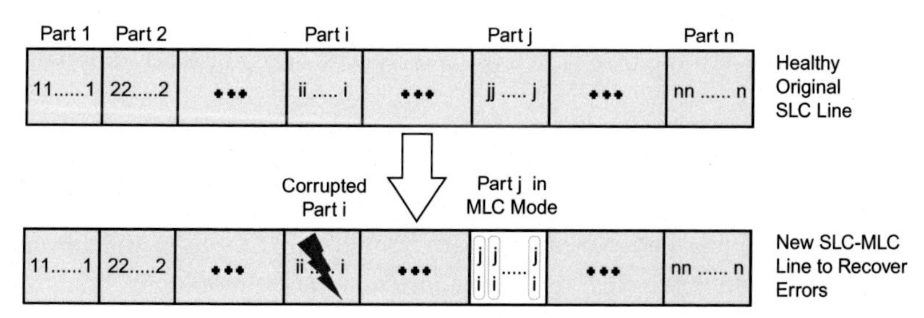

Fig. 10 Pairing a faulty part "i" with a healthy part "j"; after pairing part "i" is marked as faulty and both "i" and "j" parts are stored in 2-bit MLC mode.

Now, we assume "j" is the healthy part and can be a good target for pairing with "i." Then, both "i" and "j" have to be stored in MLC of 2 bit in the corresponding cells of part "j."

In some cases, there is more than one healthy target for pairing, so the faulty part should select the target part that is far from the faculty part in the line. This is done due to spatial locality of write requests. Indeed, if a part is written then the requests of write to the parts which are near is more probable for next coming cycles.

Our proposed method relies on determining the address of target healthy part in *healthy part selection*, and pairing process of faulty and healthy parts in *pairing mechanism*.

14.1 Healthy part selection

Healthy part selection shows that which healthy part of the line can be suitable for pairing with the given faulty part. This selection method allows the maximum freedom for choosing the target healthy part among all available healthy parts.

The target part should not have reached its endurance limit when the faulty part was paired with it. Because MLC programming further accelerates its failure and combination of the healthy part that has a short endurance with the faulty one can be counterproductive.

By checking all the parts in the line, we select the healthiest one that should not be exposed to too much write stress. Selection of pairing target is based on cell failures in that part since all parts of line can tolerate up to a limited number of cell failures. To show the faulty/healthy part, we use the valid/invalid bit. Since the number of healthy parts of a line in a page is limited, this bit can help a lot for finding the healthiest part.

Our sensitivity analysis determines that for 128B line size, 16 parts per line is enough due to large storage overhead and meta-data information. To achieve a reasonable storage overhead, better time-to-failure and performance compared to the others, we use fixed partitioning instead of dynamic partitioning.

14.2 Meta-data information

Each part in the line has two bits that are named as *MLC/SLC* and *Invalid/Valid*. To put it in more clear terms, the *invalid/valid* bit would take a state of *invalid* where the pointers of ECP in a line couldn't cover the failures of bit in a particular part. Hence, pairing of the faulty part with the part that is healthy should be done and both are stored in MLC mode. To indicate if it is programmed in MLC mode, each part would utilize a SLC/MLC bit.

There is a chance of failure to find especial healthy candidate when invalidation of all parts takes place and when all parts are MLCed in the moment a pairing is requested, so no intra-line-level pairing takes place.

We have assumed 16 parts consisting of 8 bytes for a 128B line where each partition would have 2 bits for invalid/valid bit as well as MLC/SLC bit. Therefore, ILP imposes storage overhead of 3.12%.

Although more partitions lead to greater chance for finding a part that is healthy but it comes at the cost of increasing the size of meta-data information.

14.3 Pairing algorithm

Intra-line level pairing request is invoked after finding the healthy target part by selection algorithm. Data bits of both faulty and target parts are stored and programmed in 2-bit MLC mode. In case where the MLC part fails, memory controller makes it possible to reallocate both of parts to the other healthy parts.

Read/write of MLC needs programming steps or repetitive sensing which come at cost of increased access latency or energy. Adding to it, repetitive writes have a negative impact on the lifetime, resulting in the rapid wear-out of the cell (in case of a 2-bit-MLC, each cell failure would result in two bit errors).

In practical cases, if a person selects MLC PCM, he would be bothered only about its capacity while the other parameters might be worse. The SLC PCM is, however, preferred when the person considers even the other parameters. Therefore, in ILP, when SLC level is considered, it provides the best performance, less energy and enjoys better lifetime of individual SLC cells. It also improves overall memory lifetime by using MLC scheme to make use of lifetime of healthy parts.

15. Experimental results

For the evaluation, we implemented and compared the following schemes:

- ECP-6: a system that can correct up to six errors in 128B line granularity.
- SAFER: a system that exactly models SAFER-32 scheme and corrects up to 32 errors per block.
- Free-p: a system that models Free-P scheme and can correct up to four hard errors per block.

15.1 Analysis under synthetic write traffic

The current section focuses on analysis with synthetic applications and presented parameters as it is shown in Section 8.1 in "Inter-line level schemes for handling hard errors in PCMs" by Asadinia and Sarbazi-Azad (Chapter 4).

Our results show us that while compared to that of ECP, SAFER and Free-P, the method we proposed would lead to higher memory capacity. Fig. 11 demonstrates the lifetime as compared to the memory capacity for CoV of 0.35. High process variation is evaluated which proves that the proposed method gives better lifetime and capacity as in contrast to other error correction intra-line level schemes.

Our expectation is that when we add the capability of inter-line-level pairing to this scheme, it would lead to greater capacity and lifetime of PCM. However, it wouldn't come into the focus of our current work here since we have only focused on intra-line-level error correction that is orthogonal to other inter-line-level or also a correction scheme on page-level.

15.2 Analysis under real workloads

Both multi-thread PARSEC [3] and multi-program SPECCPU2006 [9] workloads are used for our evaluation under the real workloads.

15.2.1 Performance analysis

We compared our method with FREE-P, ECP and SAFER schemes. Analysis is done under process variation with CoV of 0.25 for cell endurance. Results are shown in Table 3 and confirmed that our mechanism is lifetime and performance efficient compared to other evaluated systems. For heavy write traffic like Mix 8 and Mix 9 with more write stress, finding

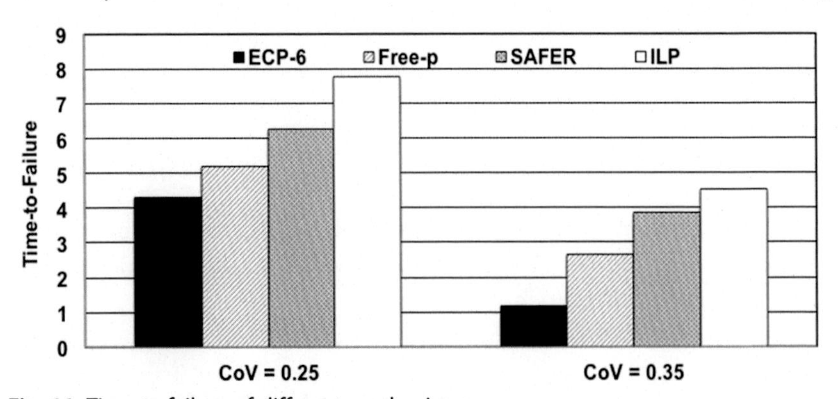

Fig. 11 Time-to-failure of different mechanisms.

Table 3 Performance analysis of ILP mechanism compared to ECP-6, Free-P, and SAFER.

Workloads	ECP-6	Free-P	SAFER	ILP
Blackschole	1	0.85	0.79	0.7
Bodytrack	1	0.67	0.72	0.55
Canneal	1	0.76	0.74	0.62
Deadup	1	0.85	0.8	0.7
Facesim	1	0.8	0.74	0.63
Ferret	1	0.82	0.75	0.66
Fluidanimate	1	0.75	0.76	0.63
Freqmine	1	0.87	0.79	0.7
Raytrace	1	0.88	0.78	0.68
Streamcluster	1	0.87	0.83	0.73
Swaptions	1	0.8	0.75	0.65
VIPs	1	0.8	0.74	0.65
X-264	1	0.78	0.75	0.64
Mix1	1	0.85	0.85	0.75
Mix2	1	0.75	0.75	0.63
Mix3	1	0.85	0.78	0.7
Mix4	1	0.75	0.75	0.65
Mix5	1	0.76	0.74	0.63
Mix6	1	0.8	0.77	0.68
Mix7	1	0.88	0.85	0.73
Mix8	1	0.85	0.8	0.74
Mix9	1	0.87	0.8	0.76
Gmean	1	0.82	0.76	0.65

The values are normalized to ECP-6 baseline system.

target healthy part is a hard procedure. Raytrace and blackscholes are examples of workloads with few writes, so the reduction in capacity has less performance impact. Mix 4 and Mix 1 are illustrations for workloads with read intensive workloads; so pairing mechanism has a less impact due to the performance bottleneck.

Table 4 Lifetime analysis of ILP mechanism compared to ECP-6, Free-P, and SAFER.

Workloads	ECP-6	Free-P	SAFER	ILP
Blackschole	1	1.2	1.5	1.7
Bodytrack	1	0.8	1.2	1.4
Canneal	1	2	2.3	2.5
Deadup	1	2	2.4	2.6
Facesim	1	1.7	2.1	2.3
Ferret	1	1.4	1.6	1.8
Fluidanimate	1	1.8	2.3	2.5
Freqmine	1	1.3	1.5	1.9
Raytrace	1	1.2	1.5	1.9
Streamcluster	1	1.8	2.2	2.5
Swaptions	1	1.4	1.6	2
VIPs	1	1.2	1.3	1.5
X-264	1	1.1	1.4	1.6
Mix1	1	2	2.3	2.7
Mix2	1	1.1	1.3	1.6
Mix3	1	1.5	1.8	2.1
Mix4	1	1	1.3	1.4
Mix5	1	1.2	1.5	1.7
Mix6	1	1.25	1.5	1.8
Mix7	1	0.7	1	1.3
Mix8	1	0.8	1	1.4
Mix9	1	0.7	1.2	1.46
Gmean	1	1.26	1.57	1.84

The values are normalized to ECP-6 baseline system.

15.2.2 Endurance analysis

Lifetime limit is described as the time which 50% of memory capacity has no scope of using within different protection mechanisms. So, we simulate each application up to that time and measure the elapsed time. Table 4 shows the endurance analysis results.

Table 5 Average number of recovered errors per line of ILP mechanism compared to ECP-6, Free-P, and SAFER.

Workloads	ECP-6	Free-P	SAFER	ILP
Blackschole	22	27	28	33
Bodytrack	17	20	21	27
Canneal	21	25	26	26
Deadup	21	27	28	29
Facesim	18	26	27	28
Ferret	22	26	27	31
Fluidanimate	17	26	27	29
Freqmine	16	26	27	30
Raytrace	15	23	24	30
Streamcluster	14	23	24	31
Swaptions	23	26	27	32
VIPs	20	27	28	32
X-264	15	27	28	32
Mix1	17	23	25	31
Mix2	13	24	26	32
Mix3	17	25	26	27
Mix4	17	25	26	33
Mix5	18	25	26	32
Mix6	24	28	30	34
Mix7	24	31	32	33
Mix8	25	30	32	35
Mix9	18	31	32	36
Gmean	20	26	27	31

15.2.3 Average number of recovered errors

The percentage of recovered cells per memory line is shown in Table 5. Compared to the other schemes, ILP can revive more cells from failures and overcome to others with higher recovery rate.

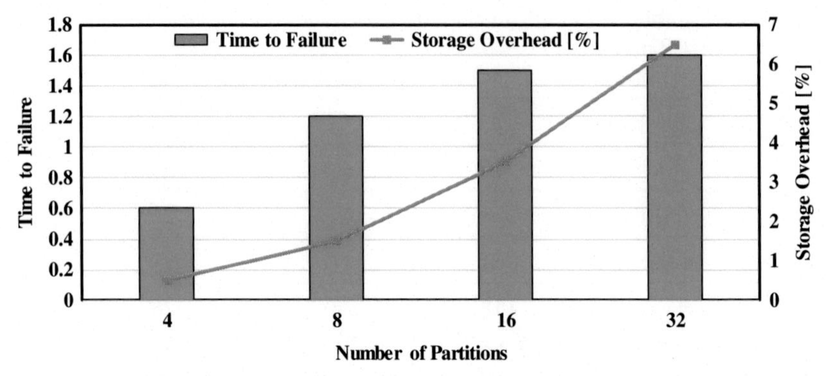

Fig. 12 Partitioning effect on lifetime and storage overhead.

15.2.4 Sensitivity analysis to partition size

Evaluation of our proposed paring partitioning size and its overhead are shown in Fig. 12. For various partitioning sizes of 4, 8, 16, 32 parts per 128B line, Fig. 12 shows time-to-failure and storage overhead. When we have less partition, finding the target healthy part for pairing is difficult since there are more parts of line that are faulty and can't be targeted for healthy parts to pair. Moreover, we have lower storage overhead but it comes with decreasing recovery as well as time to failure rate. In cases where more partitions are used, more target healthy parts are available for pairing. Although this imposes higher storage overhead and complexity of finding suitable target for pairing within each line, but it provides better memory capacity, reliability and performance. We select the partition size of 16 because there is a scope for great opportunities in finding target's healthy part with less search complexity. Besides, it provides acceptable failure coverage with an acceptable storage overhead.

16. Data block partitioning for recovering stuck-at faults in PCMs

As previously stated, write endurance limitations can lead to low cell reliability because it can cause a few memory cells to be permanently stuck at 0 or 1. Error recovery scheme is essential in solving this PCM-based memory problem and recovering from hard errors. There are a few state of the art technologies available to solve these problems at intra-line or inter-line levels. PCM endurance can be improved by remapping failed lines to spares where inter-line level schemes are considered. In addition, data block partitioning and bit inversion schemes can be used to solve issues at intra-line level. Even though the latter type methods are preferred, they might require proper data block partitioning and fault spreading across various groups. In this section, we suggest and assess an intra-line level scheme to partition a data block statically into a few groups for an efficient recovery of multi-bit struck-at faults for each partition. The efficiency is derived from a mechanism where simple shifting exploits to raise the chance of data storage in presence of failed cells [10].

Existing partitioning techniques are not capable of recovering multiple bits for each partition. Also, imposing large complexity to the memory controller in current approaches can lead to large complexity in lifetime and reliability improvement. In addition, these techniques ignore the non-uniformity of cells' activity in a data block and only consider uniformity in bit flips distribution among all cells and within memory pages.

The problem of short lifetime in the non-uniform faulty cells of data blocks makes it necessary to have a method that not only reduces the non-uniform nature of the data block but also makes the partitioning efficient. The ideal method should be less complex for the memory controller and offer a good storage overhead.

We propose an efficient data block partitioning approach. Also, it uses a simple shifting mechanism in order to recover from stuck-at faults and better utilize the failed cell with a stuck-at value because this cell is still readable. Our proposed method is highly efficient in data block partitioning. It can recover multiple stuck-at fault cells, and store the data in the failed cells.

Our proposed scheme improves the lifetime of memory during both pre and post-occurrence of hard errors. Before hard errors occur, our method uniforms bit flips distribution within each block, which results in eliminating stress from hot locations. Therefore, our proposed scheme is complementary to previous techniques such as wear-leveling methods.

When hard error occurs, there are some failed cells with permanent stuck-at faults. Our proposed method first partitions data blocks statically. Then, it shifts block contents to catch the same position with the failed cell with stuck-at value (either stuck-at "0" or stuck-at "1"). Our simple and low overhead shifting mechanism provides an error recovery technique at block-level.

Binary values such as (00000100) can be stored when cell "0" of the memory is stuck-at "1." This is doable by right shifting the data value twice (00000001) and storing the shifted version into the storage cells. This mechanism would line up the cell stuck-at "1" with the "1" in the data and both have similar position. Likewise, the associated counter of the data sets to "2" denoting the rightward shifting of data by 2. When the data is read out, both data bits of (00000001) and the counter value "2" have to be read. Then, the data will be shifted toward left to retrieve the correct data. In summary, we propose to store the shifted version of data as opposed to storing data in cells. Shifting of the data takes place such that the correct data values correspond to the values which the faulty cells are stuck-at.

17. Tolerating hard errors

This section focuses on the detailed structure of the proposed PCM architecture. We present a method to make bit flips uniform and recover hard errors. To do so, we partition a data block statically and then describe our recovery scheme regardless of the number of failed cells per partition. We use a 6-bit counter to first uniform cell usage and then for error recovery.

17.1 Before occurrence of hard errors

For static partitioning, we use 8 partitions with 8 bytes each. We also use a 6-bit counter to make the bit flips uniform and implement our shifting scheme. Upon receiving a write request, the block's content is shifted by 1 bit for each block. This effective and low overhead approach can prevent non-uniformity in bit flips across memory blocks, remove stress from hot locations and relax the worst-case bit flip rate in each line. Fig. 13 illustrates the partitioning and shifting mechanism of our proposed scheme.

17.2 After occurrence of hard errors

After each write, memory controller would perform an additional read to determine whether the write operations are successful or failed. This approach is efficient because latency and energy of read operations in

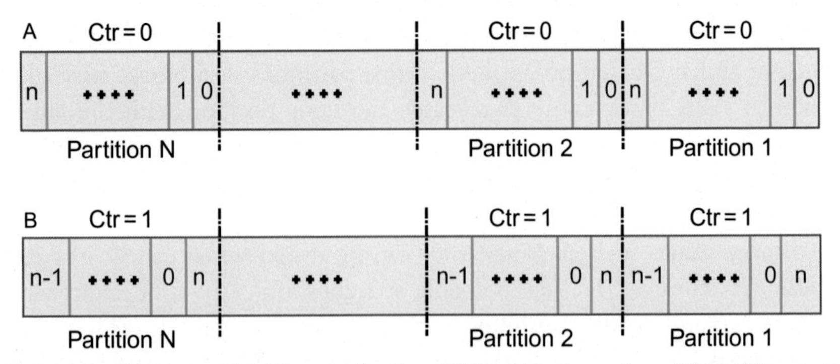

Fig. 13 Partitioning and shifting mechanism. (A) Data block consists of N partitions and each partition has its own counter. (B) For each write, the content of each partition is shifted by 1-bit.

PCMs are more cost-effective than writes. Hence, our method detects errors by performing read-after-write without causing performance loss. This fact has been established and practiced by the community following non-volatile research.

During hard error occurrence, block content is shifted until the shifted data fit in the stuck at cells. This formation can be observed when a limited amount of shift is required as a result of high information redundancy and finding frequent value locality. Then, based on the shift count, our recovery mechanism can recover data by using the correct value from the shifted data block. If we cannot find the matching state due to the specific data pattern, a permanent fault occurs that would make the entire memory block useless for the memory manager in the future.

17.3 Examples of the proposed method

Fig. 14 illustrates the operation and organization of our proposal. Fig. 14A shows how data block splits into N partitions and each partition has its own counter. Fig. 14B represents the state before hard error occurrence. Upon receiving each write, the content is rotated by 1 bit and simultaneously updated. This action of rotation helps in the pressure removal of bit flips when we see a particular block cell.

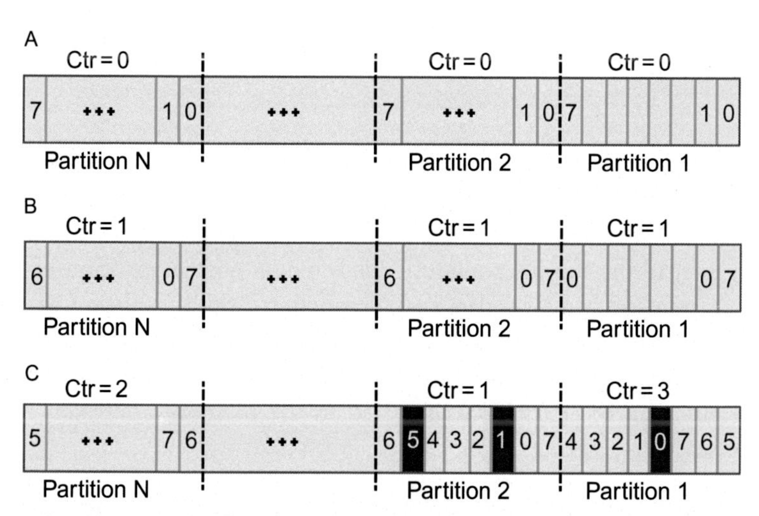

Fig. 14 Partitioning and shifting example: (A) Static partitioning and counter of each partition. (B) The content is shifted by 1-bit in order to get uniformity upon receiving a write then the counter is increased. (C) Third position of first partition and the fourth place of second partition suffer from stuck-at faults. In this case, the shift counter is used. The content is rotated and the right place which is compatible to our faulty cells is found.

When hard error occurs as represented in Fig. 14C, faulty cells exist in both partition 1 and partition 2. So, we rotate the partition 1's content by 3 bits for a proper mapping of data on faulty cells. Partition 2 follows the same approach, but shifting would be done only by one bit in order to cover the stuck-at cell. While partition 3 doesn't have any error, we would shift one bit for better bit flip distribution.

18. Write operation

We consider two possible scenarios for writing a data block: (1) before occurrence of hard errors, and (2) after occurrence of hard errors. If there are no hard errors, then the block content is shifted and written the values. We should highlight that shifting is based on the value determined by corresponding counter. During the write request to the memory controller and verification of unsuccessful write by the checker read, mask bit is created. Our method can realize the number of required shifts based on the created mask bit which displays faulty positions in the data block. This means having faulty cell positions in each partition can help shift the data block such that shifted form has similar data with the value of faulty stuck-at cells. A desirable state would have a maximum of eight times shift as per sensitivity analysis in the evaluation part. If we can find such a state, then a new write that has a refined data gets issued to perform this pending write. Otherwise, this page is discarded from OS space and write should get issued to the new allocated page.

19. Read operation

When a read request is in the process, memory controller sets the read circuit to read the block's content. This is why a normal iteration of SLC read gets released to deliver data to the requester. After this, the correctness of the content can be achieved based on its shifted form. With regard to the counter value, each partition is shifted to put every bit in its original position. In the last step, the data block is passed to the memory controller to be delivered to the read requester.

20. Meta-data information

We use 64B line with 8 partitions (each for 8 bytes) and a 6-bit counter for each partition. Counter is used to implement our shift-based mechanism and to reduce average bit flip pressure. In presence of hard error occurrence, the counter is used to indicate the number of shifts applied to the block content. Therefore, per data block, we require $6 \times 8 = 48$ additional bits which results in $48/512 = 9\%$ storage overhead.

21. Experimental results

To assess our method, we use the simulation tool and simulation methodology presented in Sections 7 and 8. We then compare our proposal to the stat-of-the art hard error recovery methods (ECP-6 and SAFER-32).

21.1 Lifetime

Specific writes can cause a loss in the memory capacity by 50%. The elapsed time until the 50% reduction in memory capacity is defined as lifetime duration. In Fig. 15, the curves represent the lifetime for various schemes. ECP-6 and SAFER-32 are compared with the proposed scheme, where our scheme promises the improvement in a lifetime by up to 35% and 22%, respectively. This is by reusing the failed cells in our partitioning scheme. This can be achieved by using a simple and efficient shifting mechanism.

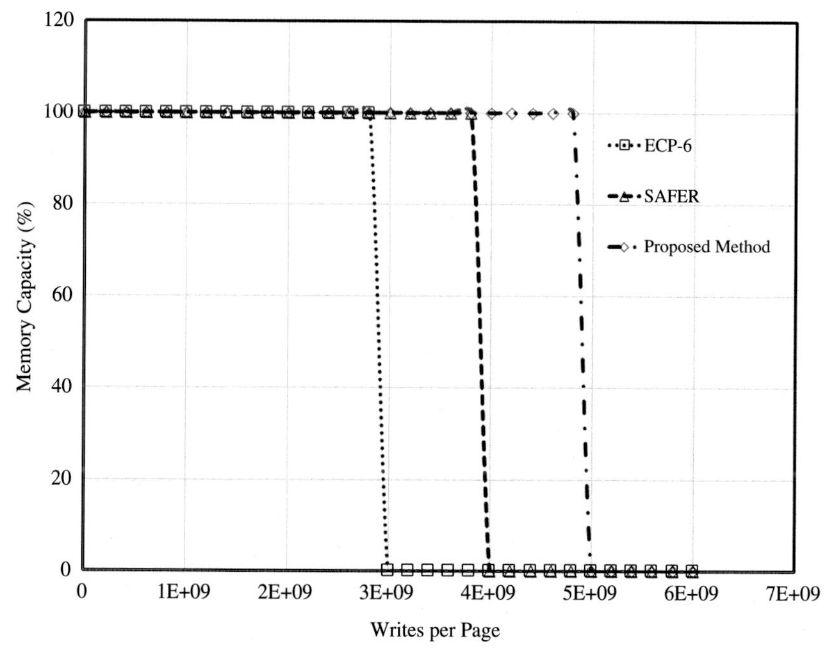

Fig. 15 Capacity degradation of the proposed method compared to ECP-6 and SAFER-32.

21.2 Average number of recovered errors per block

Our recovery mechanism has a higher recovery rate compared to other schemes, as shown in Fig. 16. This improvement relies on reusing faulty cells and recovering multi-cell failures in each partition.

21.3 Partitioning size effect

Fig. 17 illustrates different partitioning sizes and their impact on the storage overhead and efficiency of the proposed scheme. For limited partitions, the requirement of bits per block is less as a result of less number of partitions. For instance, 4 partitions require multiple of 7 bits that are 28 bits per line. Though the probability of failure occurrence in each partition is high, our recovery mechanism is less efficient due to hard matching between the data pattern and the faulty cell. But when we use more partitions, more bits per block are needed. This leads to a larger meta-data storage overhead. But this

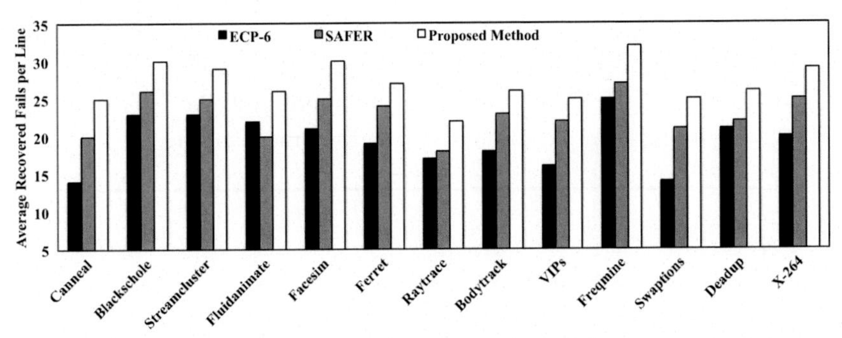

Fig. 16 Average number of recovered errors per line of the proposed mechanism compared to ECP-6 and SAFER-32.

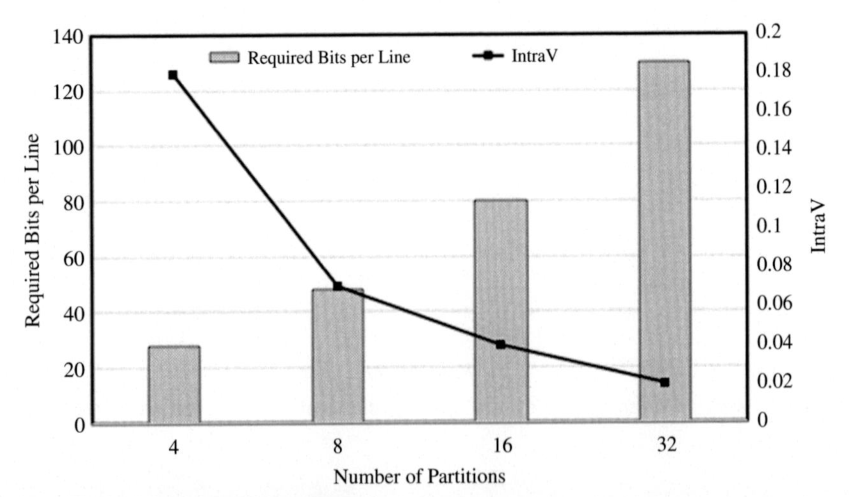

Fig. 17 The effects of partitioning.

has higher opportunity to handle failures in various partitions. When the partitioning size is eight, we have a good trade-off between the reliability issues and acceptable storage overhead.

21.4 Overall performance

In our observation, Cycle per Instruction (CPI) is considered as the overall performance metric. In Fig. 18, our mechanism has improvement as compared to the ECP-6 and SAFER-32. This is a result of longer lifetime of memory blocks in our method and less accesses to the external SSD.

21.5 Number of required shifts

Fig. 19 shows the distribution pattern of required shifts to find a position that is suitable. But as can be seen in this figure, the workloads that are multithreaded use up to 11 shifts when the average is only 6 shifts. It is important to mention that multiple shifted patterns (here 0–11 shifts) have to be done

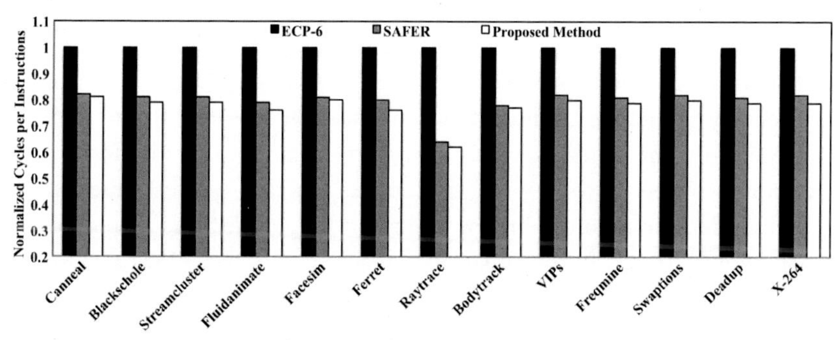

Fig. 18 Performance analysis of the proposed scheme compared to ECP-6 and SAFER-32. The values are normalized to the ECP-6 as the baseline system.

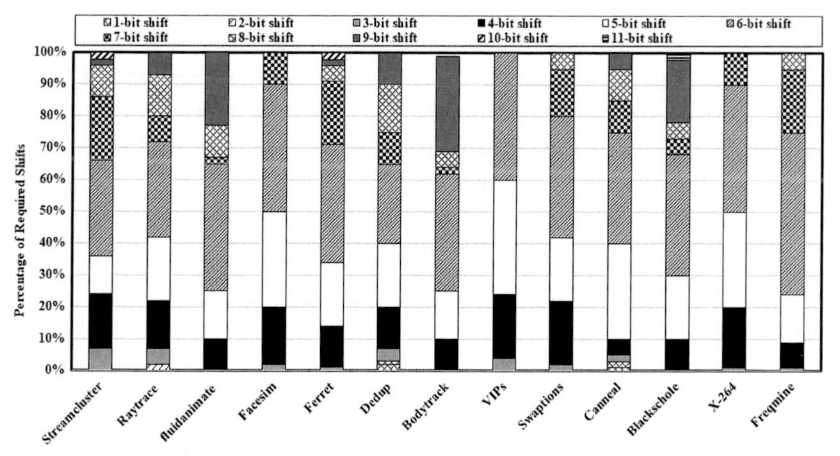

Fig. 19 Average number of required shifts per line.

in parallel. Fig. 19 shows that the 8-bit shift could be highly efficient in finding a suitable position and that would lead to a longer lifetime. Shifting is more efficient when it uniforms bit flips like wear-leveling scheme and is less efficient when it leads to increase in bit flips. However, our experimental results confirm that the positive impacts can outweigh the negative impacts due to the data locality and similarity. Therefore, we highly recommend the shifting mechanism.

21.6 IntraV

We use IntraV metric to show the reduction in bit flips' variation for different multi-threaded workloads. This measure is reported with and without applying our scheme. We observed that we have about 20% reduction in bit flips' variation in each block.

21.7 Latency overhead

Our proposed method highly relies on the shifting mechanism. But, when a read or write operation takes place, we don't operate on a longer row size for the memory to include counter values for each line. Instead, counter values are stored in some other alternative memory lines. So, the additional read operation is necessary to retrieve the counter values. Counter values could be included in the memory line by making it longer with the help of additional memory blank. In this situation, a whole line of data block and metadata counter values need to be read or written in one particular memory cycle. Clearly, memory lines that store counter values seem to impose higher latency overhead due to additional read operations required to receive counter values. Alternatively, we can use caching mechanism to reduce the negative impacts of this arrangement. Or caching mechanism is similar to the failed cache used in the past proposals. It decreases the latency overhead due to additional read operation to receive counter values. Hence, a fraction of meta-data is kept in the small cache of the memory controller. This makes sure that a major part of memory accesses to counter values could be shown in the small cache making reduction in latency overhead.

For example: 48 bits overhead for each block requires a 2 KB cache to store meta-data for 341 blocks (2048/6 = 341). Therefore, we evaluate our proposed technique by both using and not using small failed cache. The performance result is shown in Fig. 20 and it demonstrates higher gain in performance when we use small failed-cache.

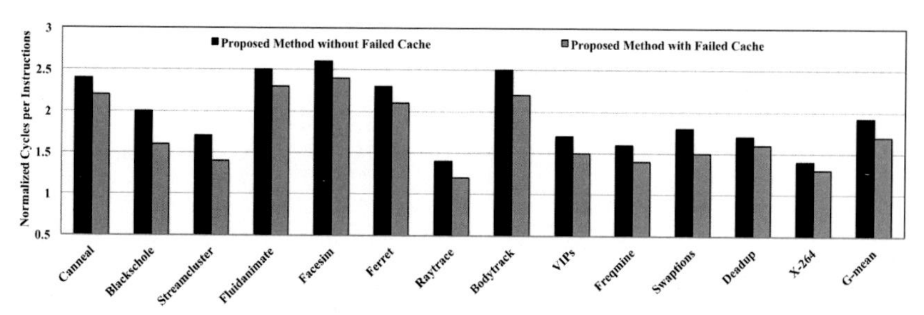

Fig. 20 Performance analysis of the proposed scheme with and without using the small failed-cache.

References

[1] Asadinia M, Jalili M, Sarbazi-Azad H: BLESS: a simple and efficient scheme for prolonging PCM lifetime. In *Proceeding of Design Automation Conference (DAC)*, 2016, pp 1–6.

[2] Binkert N, et al: The gem5 simulator, *ACM SIGARCH Comput Archit News* 39:1–7, 2011.

[3] Bienia C, Li K: PARSEC 2.0: a new benchmark suite for chip-multiprocessors. In *Proceedings of the Workshop on Modeling, Benchmarking and Simulation (MoBS)*. 2009.

[4] Schechter SE, et al: Use ECP, not ECC, for hard failures in resistive memories. In *Proceedings of the International Symposium on Computer Architecture (ISCA)*. 2010, pp 141–152.

[5] Wang J, Dong X, Xie Y, Jouppi N: i2wap: improving non-volatile cache lifetime by reducing inter-and intra-set write variations. In *IEEE 19th International Symposium on High Performance Computer Architecture (HPCA'13)*, 2013, pp 234–245.

[6] Ipek E, Condit J, Nightingale EB, Burger D, Moscibroda T: Dynamically replicated memory: building reliable systems from nanoscale resistive memories. In *ASPLOS*, 2010, pp 3–14.

[7] Qureshi MK: Pay-as-you-go: low-overhead hard-error correction for phase change memories. In *Proceeding of the IEEE/ACM International Symposium on Microarchitecture (MICRO)*, 2011, pp 318–328.

[8] Asadinia M, Sarbazi-Azad H: Using intra-line level pairing for graceful degradation support in PCMs. In *IEEE Computer Society Annual Symposium on VLSI (ISVLSI)*. 2015, pp 527–532.

[9] Spradling CD: SPEC CPU2006 benchmark tools, *ACM SIGARCH Comput Architure News* 35(1):130–134, 2007.

[10] Asadinia M, Jalili M, Sarbazi-Azad H: Data block partitioning for recovering stuck-at faults in PCMs. In *International Conference on Networking, Architecture, and Storage (NAS)*. 2017, pp 1–8.

CHAPTER SIX

Addressing issues with MLC phase-change memory

Marjan Asadinia[a], Hamid Sarbazi-Azad[b]
[a]University of Arkansas, Fayetteville, AR, United States
[b]Sharif University of Technology and Institute for Research in Fundamental Sciences (IPM), Tehran, Iran

Contents

Abstract

All of the presented solutions in this book focused on using MLC phase change memory (PCM) due to density advantage and prolonging PCM lifetime. However, *resistance drift* can be one of the challenging issues for MLC PCMs. While it is desired to have the density advantage of MLC, the trade-off is resistance drift. Since MLCs have closely separated resistance regions, drift has a chance of overlapping intermediate regions. It may then bring out either single bit or multi-bit soft error. Indeed, drift source is related to the semi amorphous resistance regions that are metastable vs time and temperature while crystalline resistance proves to be stable across time and temperature. This chapter solves this challenge of resistance drift problems by designing a

Advances in Computers, Volume 118
ISSN 0065-2458
https://doi.org/10.1016/bs.adcom.2019.10.004

memory architecture and circuit. It improves energy, latency and reliability of MLC PCM while maintaining its capacity advantage. The solution we offer is variable resistance spectrum MLC PCM.

1. Variable resistance spectrum assignment

The variable resistance spectrum MLC phase change memory (PCM), or VR-PCM in short, is developed after observing that the wide range resistance spectrum of GST promises to configure resulting in non-uniform regions with each having different read/write latency and energy [1]. What the observations prove is that a feature of GST resistance regions in the VR-PCM is that they are either shrunk or enlarged. This is distinct to conventional MLC PCM that has uniform partitioning. This distinction of VR-PCM ensures that energy dissipation and access latency of various regions are unequal. It means that at complete amorphous side, access to the most enlarged partition offers a single cycle, unlike the complete crystalline side where the most shrunk partition offers a little larger access latency compared to a conventional MLC. Therefore, results of VR-PCM are phenomenal since it provides low energy consumption with higher PCM endurance. This is because on the crystalline side, VR-PCM reduces bit error rate through narrowing the interstate noise margin. In VR-PCM, access latency and energy consumption can be reduced because of the unbalanced read/write access latency which is fitted well to the non-uniform data pattern distribution of memory transactions (known as frequent value locality FVL). The data replication decreases while bit storage level of a single cell increases. However, for 3 bit or 4 bit MLCs, which are known as high-density devices, VR-PCM can have negative impact on main memory's desired features like high bandwidth, high endurance, low latency and low energy.

Therefore, negative side effects of VR-PCM can be reduced by on demand non-uniform partitioning during program execution phases. Steps to mitigate any negative side effects of VR-PCM are as follows, first, determining resistance regions of various pattern in a fair manner through a conservative binary directed, non-uniform resistance partitioning scheme. Second, deciding the use of uniform or non-uniform resistance partitioning by using a monitoring scheme along with reconfigurable access circuit.

2. The MLC VR-PCM

In this section, details of VR-PCM architecture and the way it improves MLC operations are described.

2.1 VR-PCM for MLC read improvement

Maximum read latency along with the number of reference cells in conventional PCMs can be minimized by sequential bit-by-bit data readout. On the positive side, sequential bit-by-bit provides an optimization opportunity that VR-PCM can enhance energy efficiency and latency of MLC read by intelligent use of it as a trade-off for memory controller complexity and overhead of reference cells. However, it leads to equal latency and also energy cost for all the states. So, we introduce and utilize the feature of "*Frequent Value Locality*."

Within a memory system, a limited set of distinct values result in a large fraction of fetched or stored data based on frequent value locality [2]. For example, PARSEC-2 benchmark [3] was examined for this phenomenon through capturing main memory traces from full-system simulation.

The illustration in Fig. 1 shows that the accessed values frequency observed from high to low at 4-bit granularity. Based on this figure, 75% of the contribution of four MFV or four most frequent values is exhibited from five left most programs while 65–75% of such values are contained in next four programs along with 65% MFV in last three programs.

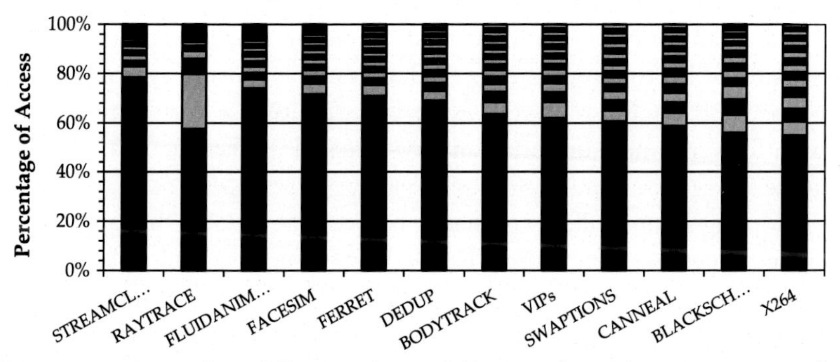

Fig. 1 Percentage of 16 different values read/written from/into main memory for PARSEC-2 benchmark at 4-bit granularity (from most to least).

In VR-PCM design, a sequential readout scheme is used along with increasing the reference cell numbers to $2^N - 1$ which will result in reduced read latency, lower energy consumption of an N-bit MLC memory without sacrificing its density.

In RDC, the value differentiating state i from state $i+1$ (where $i = 2^N$ for crystalline and $i = 1$ for amorphous) is represented in the value of ith dummy cell. Indeed, the S/H output is first compared with RDC_1 during a read operation to check if the resistance values of read operation lies within DP_1, resistance region coded completely as amorphous (resistance region of data pattern 1). However, if RDC_1 is more than read resistance, to read data value DP_2 in second cycle, S/H output is compared with RDC_2 iteratively. It is then continued to compare it with RDC_3 in the third cycle for reading the data value DP_3 and so on.

When the stored value is DP_i, the above pattern by pattern read mechanism takes i cycles. Thus, this method shows less latency for reading the first k patterns which is coded by the amorphous side phases that is ($k \leq \log_2 N$) while the latency is larger for the remaining $2^N - k$ values on the crystalline side.

Compared to conventional bit-by-bit scheme where the latency is constant, equaling to $\log_2 N$ cycles, VR-PCM provides improvements in read latency and energy if these k patterns are the first k most frequent values.

2.2 Binary-directed resistance partitioning

In real systems where high degree of value locality is achieved from real workloads, we can use our proposed VR-PCM since it provides even single or few limited cycles for MLC read operation. MLC cell reliability and write latency has to be enhanced which is desirable for such systems, so our proposed VR-PCM solves this issue by exploiting a low complex and straightforward non-uniform binary-directed resistance partitioning scheme.

Adopting a non-uniform resistance partitioning of GST that gives more weights to regions seeing more drifts can help to improve readout MLC reliability and also relax the repetitive tendency of the PCM writes. It means that the semi-amorphous states of highly resistive regions are enlarged and follow the uniform baseline partitioning as shown in Fig. 2; however, complete crystalline regions are quite shrunk.

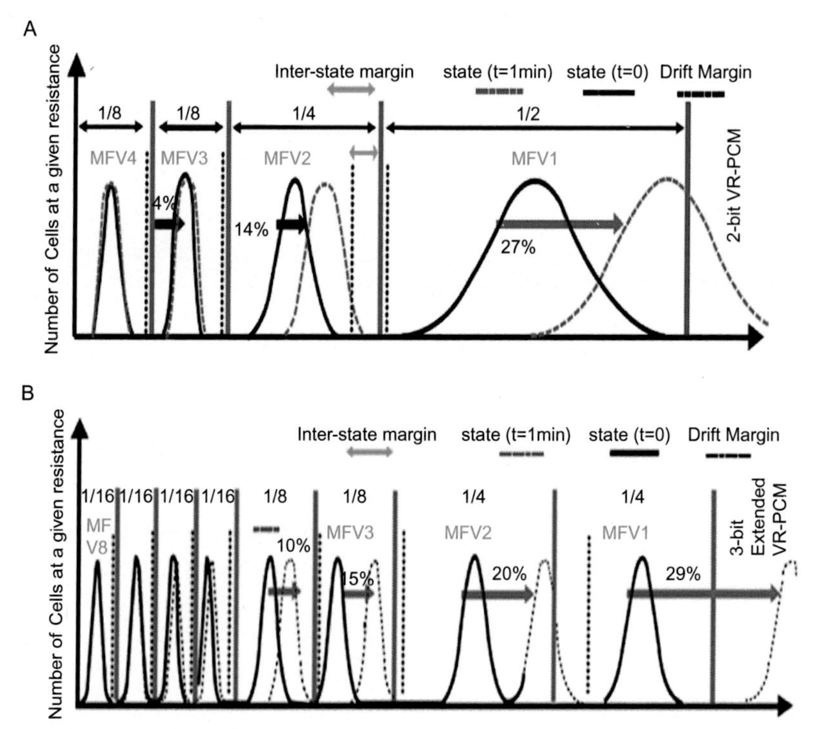

Fig. 2 (A) VR-PCM binary resistance partitioning for 2-bit MLC. (B) Extended VR-PCM partitioning for 3-bit MLC.

An illustration of non-uniform exponential halving rule by 2-bit VR-PCM is presented in Fig. 2A top. It shows an example of resistance spectrum division:

$$\overbrace{1/2, 1/4, \ldots, (1/2)^{N-1}}^{(N-1)\text{amorphous side}}, \underbrace{\frac{(1/2)^{N-1}}{2^{N-1}}, \ldots, \frac{(1/2)^{N-1}}{2^{N-1}}}_{2^N-(N-1)\text{Crystalline side}}$$

BDRP or binary-directed resistance partitioning is a simple and also straightforward process that shows that the number of reference cells for the sense amplifier (SA) reduces to $N-1+2^{N-1}$.

2.3 Discussion

Below focuses on new opportunities that are given by configuring GST resistance spectrum based on BDRP rule.

2.3.1 Enhanced write latency, energy, and cyclic endurance

VR-PCM reduces the number of iterations for the stretched region as SET pulses which are required in P&V process clearly shows a strong correlation with distribution of resistance in target range. However, this is not true for shrunk regions where there is increase in maximum number of iterations.

Based on the target resistance range, we can assign patterns of data to the regions which are stretched either using selective RESET-to-SET or SET-to-RESET P&V methods [4] or we can use top elements of frequent value stack to deal with the problem of energy/endurance/performance loss of writes.

To improve write energy and cyclic MLC PCM endurance, our observation (evaluation in the next section) proves that reducing P&V iterations can be utilized. It also helps to improve latency since there is strong correlation between value locality and spatial locality of memory elements.

2.3.2 Improved retention time and drift tolerance

Retention problem is defined in such a way that one stored binary state turns into the other state due to either capacitance discharge in DRAMs, PCMs Phase changing (i.e., changing from amorphous to crystalline or vice versa) or magnetic based mediums such as MRAM getting demagnetized. This is why retention period has a close relationship to how much the device can tolerate the degradation of binary state before the data that is stored is erroneously converted. As interstate noise margins are narrower in MLC PCMs, resistance drift has potential to substantially reduce retention time. This shows that resistance drift highly depends on retention time and they are not separate issues.

In this line, the more vulnerability to drift exists in the narrow resistance bands that result in lower retention time. However, our VR-PCM design offers maximum drift tolerance in the narrowest regions (those regions near to crystalline state). This promises reduction in data loss probability due to drift. On the other hand, widened regions (those near to amorphous state) can tolerate more drift.

To summarize, as drift tolerance is improved in VR-PCM, it also offers longer retention period along with data integrity which rarely require refreshing mechanism [5]. Moreover, VR-PCM provides a lower probability of MLC PCM erroneous readout with narrower interstate margins.

2.3.3 VR-PCM shortcoming in high-density MLC

MLCs with high storage density in VR-PCM structure impose super-linear increase of the maximum read/write latency. Moreover, efficiency of VR-PCM depends on the replication rate of the few first MFVs which then diminishes for large data storage like 3-bit and 4-bit. VR-PCM partitioning has a chance to be inefficient where high-density MLC is involved due to the non-uniformity allocation of resistance bands. To overcome the VR-PCM's negative side effects, the solution lies in allocating non-uniform partitioning on demand and also during program phases that can derive benefit from it. To reach this goal, the below sections discuss two approaches which include concepts of different designs.

3. Extended VR-PCM

For an N-bit MLC, the first step is to have exponential halving rule for resistance partitioning of an (N − 1)-bit VR-PCM. As illustrated in Fig. 2, second step includes doubling the number of resistance regions (model 2^N distinct regions in an N-bit MLC) by dividing any region into two separate equal-sized regions with a noise margin, which then separates these two equal regions. This new structure is termed "*Extended VR-PCM*" where minimum read/write latency is doubled and then shown as a large portion of VR-PCM performance loss mitigation due to enhanced maximum latency.

Compared to VR-PCM, reliability level of drift in the new Extended VR-PCM is compromised for fast and low energy accesses to less frequent values. Although using wider noise margins in this design can improve lower drift tolerance, it comes at a cost of lower cell lifetime and slightly higher write energy.

4. Ultimate design: Reconfigurable VR-PCM

Extended VR-PCM no doubt is applicable for high-density MLCs and appears to be more productive than VR-PCM, but it still suffers from VR-PCM shortcomings like serious latency/energy problems as all worst-case resistance regions (most shrunk ones) play dominant roles. The solution is that we benefit from the advantage of reconfigurable PCM access circuit which can alter between both conventional uniform resistance partitioning mechanism with bit-by-bit inspection mechanism (where frequent value locality having low coverage) and also non-uniform Extended VR-PCM resistance partitioning with a pattern-by-pattern read mechanism (where frequent value locality has high coverage).

This suggestion/idea thereby involves in allocating non-uniform partitions on demand and also during program phases. Hence, when low value replication rate reduces VR-PCM efficiency, PCM memory's performance or energy of the VR-PCM memory will not be badly affected.

During data block readout, to keep the consistency of a previously written data block, we embed a 'VR bit' to each memory line with size of the last-level cache line. The VR bit is 1 when the stored data is in VR-PCM mode, so we follow the binary-directed resistance partitioning. However, we have to follow the uniform partitioning when the stored data is in conventional MLC PCM and VR bit shows 0.

In fact, when read operation is in process, the VR bit is read and examined by memory controller before main memory access.

To access to VR-PCM or conventional MLC mode, memory controller performs reference cell set selection or reference inputs activation of sense amplifier based on the output.

Therefore, cell contents are accessed by reconfigurable VR-PCM through metadata information, which includes VR bits. VR Table is a lookup-based storage and a unified structure. It can store meta-data information. It is activated and then involved in selecting by using the same word line of the corresponding memory line. The size of the VR Table is dependent on memory and cache line size.

To illustrate, assume we have the 4GB 2-bit MLC PCM main memory with 128B block size. It can be utilized to use VR Table of size 32MB. Hence, if stored in a fast DRAM array, it imposes 0.78% capacity overhead and 3.12% area overhead with the formula of ($\text{Area}_{\text{2-bit MLC PCM}} = 4 \times \text{Area}_{\text{DRAM}}$) [6,7].

5. Hardware implementation issues

In this section we discuss VR-PCM hardware implementation. First, we would prove the function of PCM controller that is modified to achieve binary-directed resistance partitioning. Then, we show the theory behind the logic of frequent value finder. Finally, we discuss VR-PCM accuracy and efficiency targeted by certain operating systems.

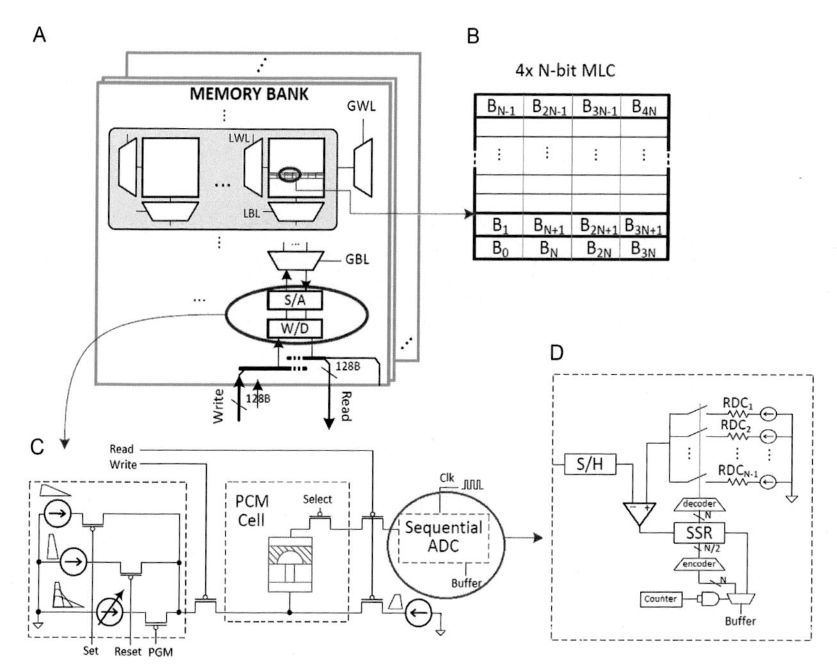

Fig. 3 VR-PCM array structure and its controller.

5.1 Main memory controller

Fig. 3A illustrates the array structure of VR-PCM and confirms its similarity as that of MLC PCM commodity designs with the controller being slightly modified. Adding to it, VR-PCM consists of hierarchical organization of multiple banks that has write driver circuits along with shared sense amplifier. It also includes blocks and sub-blocks which individually has private column and row decoders.

Furthermore, as shown in Fig. 3B, data contents are tightly vertically stored in MLC devices. Although VR-PCM array architecture is similar to conventional MLC PCM, specific design issue of it has to be carefully investigated.

Fig. 3C illustrates sensing mechanism of VR-PCM that consists of a priority encoder along with a different voltage comparator. RDC's voltage reference is compared to the cell resistance at S/H output. Voltage sense amplifier (SA) takes care of this comparison and its output stores in a serial shift register (SSR). For the next iteration, SSR selects reference voltage. More precisely, at beginning of $N-1$ iterations, and when SA output is 0,

SSR content is first transmitted to priority encoder to find out the representative patterns of data (i.e., pattern-by-pattern read mechanism). Later, if there are any remaining $2^N-(N-1)$ iterations, SSR output which uses bit-by-bit readout scheme, specifies the output. It is notable that controller uses a $\lceil \log_2(N-1) \rceil$-bit counter to track the number of iterations and determine the valid output (by priority encoder or SSR). To support VR-PCM requirements, the access circuit of the baseline is modeled in HSPICE with 45-nm technology node and slightly changed it. Compared to the conventional PCM read circuit, we achieved 0.7 ns and 0.36 pJ (<1%) for read latency and energy overhead of the modeled access circuit.

5.2 Frequent value finder logic

For value locality utilization, there is a necessity to propose a fast method to quickly identify the relative occurrence frequency of data patterns. Stable and similar relative frequency is achieved in single-thread programs for different values. To derive frequent values, a monitoring scheme like the one is presented in [2] can be used during starting phases of the program. It should be noted that frequent values extract from multi-program workload as they provide variable locality ratios at various runtime intervals and they are highly dynamic. In our observation, we also noticed that in multi-program workloads, there are two categories where most frequent values can divide. One is where very frequent values are limited. This type is very common across different applications especially all-zero and all-one patterns in most cases. There is, however, a large set of other MFVs, that alter or differ from application to application. What differentiates latter can be identified through a monitoring scheme.

There are two tables, each separately in finding logic. First table, which is FIFO buffer, uses during different phases and can filter transient non-frequent patterns occurrence. Here in this table, each entry has a *value* field that stores the data value and also *saturation counter* field, which tries to track data transfers frequency. The CAM table would be searched whenever data pattern occurs in a transaction. A hit is happened when the data is found and this finding results in increasing the corresponding counter. It then considers as an frequent value candidate on counter's saturation. If it is not found (a miss), all counters of this table would be decremented by one while in cases of those that are less than a threshold, existing new data pattern replaces with previous entry.

Frequent value table of small-sized, however, is used for storing both the code word and value of the frequent patterns. In this table, each entry would have a *counter* field, a *pointer* field, a *value* field and a used bit. The counter field basically is a saturating one which would be incremented on every access of frequent value. Pointer field would involve in counting the number of cache blocks that are coded by this value pattern. If a line in this table doesn't refer to any frequent value where used bit is 0, so it is marked as *Gap*. However, a line in the *frequent value table* with a counter value less than the threshold is considered as a victim if a value is marked as a frequent pattern in the FIFO buffer and the Gap line in the frequent table is free (used bit = 0).

When a victim line would be marked as Gap with *used* bit set to 1, from *FIFO buffer*, a candidate value is stored in the *frequent value table* (with pointer fields and zero counter).

The transferring of memory blocks in and out results in decreasing the pointer field of the Gap line. When the pointer field becomes zero, its used bit is reset. It is notable that the *frequent value table*'s counter field should be longer than that of the *FIFO* table. Indeed, it is required for operating the frequent value finding logic efficiently.

The productivity of frequent value finder logic can be studied by considering nine 4-application multi-program SPECCPU2006 workloads [8]. So, we split the program into multiple monitoring steps each with 100 million SPARC instructions.

During each monitoring step, ideal frequent value set which is obtained by software monitoring of the collected traces is compared with the collected frequent values of memory accesses. The proposed frequent value finder logic has high accuracy and effectiveness for finding the similarity between frequent value accesses.

5.3 Reassigning resistance levels to data values

During a single program, the stability of frequent data values is ensured. So, it is cheaper to take help of a monitoring scheme at the beginning of the program such as initial 5% [9] to derive frequent values and rely on the derived MFVs. This is the phase of identification, after this phase the problem of reassigning the level of resistance in data that already exists in memory system arises. The solution is to first read out and later rewrite every data that exists. This is not usually recommended due to the high cost at which it comes in terms of energy, wear out and performance.

To achieve this goal in the reconfigurable VR-PCM structure, VR bit is used and when data value is stored during frequent value detection period, VR bit of a memory line is set to 0. Moreover, it is set to 1 in the case where data value is stored during VR-PCM's normal operation. When FV detection period is over, the memory access circuit is reconfigured for bit-by-bit readout scheme if a line (with a VR bit = 0) is going to be read. On the other hand, if a line is going to be written, its VR bit is set to 1 and VR-PCM policy is followed to store data pattern.

It leads to avoiding extra cost for energy, latency and lifetime in the earlier data blocks that are stored by using a reconfigurable access circuit. Reconfiguration of PCM memory can be dealt through VR bit when there is a need to alter between the non-uniform and uniform resistance partitioning.

5.4 OS support for value translation

When application requirements and data patterns within an application change at runtime, we need to discuss the operating system support for our proposed VR-PCM method. The important feature of VR-PCM mechanism involves frequent data values identification. However, the most frequent values will be totally different for various workloads in multi-tasking and multi-core processors operating systems. To support this difference in most frequent values, there is a need for mechanism of workload-based value translation.

For example, we have 011 for the first most frequent value MFV_1 of workload$_1$ and 001 for the MFV_1 of workload$_2$. In our mechanism, both MFV_1s would be allocated to the highest resistance level, an equivalent MLC cell resistance level. Therefore, there is a need for translation mechanism.

To solve this, we tend to assume that translation of value is completed at OS level employing a small look-up table. Such associate OS-supported PCM main memory management would ease translation of value once running multiple applications. Indeed, in page and segment-based memory management, the owner of each physical memory page is known by OS (using limit and base registers).

VR-PCM management process can use this privacy at operating system level for checking the value translation table in which resistance regions assignment for encoded data values are stored per program. On the other hand, in presence of the shared and sparse data blocks, conventional uniform resistance partitioning is used for data encoding by OS and OS tags these pages as normal with VR bit = 0.

6. Simulation results

Here we tend to provide the results of our full-system simulation where VR-PCM is compared to the traditional design of MLC PCM as a baseline. For analysis, we use the configuration of system, logic model and workloads, of the PCM memory represented in "Inter-line level schemes for handling hard errors in PCMs" by Asadinia and Sarbazi-Azad (Chapter 4, Section 4.7).

Coming to concept of interstate noise margins, we decide to use the prototype model as described in [6,10] where the complete crystalline and complete amorphous states of PCM devices would be equal to 10^6 and 10^3 Ω, respectively. We also consider the noise margin that can tolerate drifts for about 1.2 s at 300 K for the baseline. So, the width of the interstate margin for MLC devices follows the rule of X, 2X, 4X,... with the value 10 KΩ for X in 2-bit MLC, 5 KΩ for X in 3-bit MLC, and 3 KΩ for X in 4-bit MLC.

In VR-PCM, the states close to complete crystalline need narrower noise margins. So, the noise margin of interstates for VR-PCM is set in such a way that it can tolerate resistance drift for a maximum of 1.23 s at 300 K. In the upcoming section, we analyze the soft error reliability of VR-PCM for the evaluated system with 4 GHz frequency.

The non-uniform division in the VR-PCM highly affects parameters of write/read latency/energy. For write, we use completely RESET and completely SET pulses for the data patterns stored in amorphous and crystalline states, respectively. In this line, we use the two-phase model which is validated in [11] to find out the number of P&V steps.

Learning phase for the primary few iterations along with a practice phase is contained in the model to calculate the probability of finishing MLC write at iteration k with Bernoulli distribution as:

$$P(k) = \begin{cases} F_1 \cdot (1 - F_1)^{k-1}, & \text{if } k \leq i \\ F_2 \cdot (1 - F_2)^{k-i-1} \cdot (1 - F_1)^{k-1}, & \text{if } k \geq i \end{cases}$$

where F1 and F_2 would be expected probabilities of convergence in *kth* iteration that continues for each phase; the number of iterations in the learning phase is i, while k denotes the overall number of P&V steps.

6.1 VR-PCM for 2-bit MLCs

When we use binary-directed resistance partitioning, we expect large improvement for 2-bit MLC PCM. The low density (2-bit) VR-PCM results for write/read access latency can be seen in Table 1. A same level of cell density is taken to normalize the results to conventional MLC designs.

Table 1 Read and write access latency of 2-bit VR-PCM normalized to the conventional MLC design in the evaluated 4-core system.

Workloads	Read access	Write access
Blackschole	0.8	0.5
Bodytrack	0.9	0.5
Canneal	0.85	0.45
Deadup	0.78	0.6
Facesim	0.8	0.6
Ferret	0.8	0.5
Fluidanimate	0.7	0.6
Freqmine	0.85	0.7
Raytrace	0.78	0.1
Streamcluster	0.65	0.5
Swaptions	0.8	0.48
VIPs	0.63	0.8
X-264	0.82	0.57
MP-H1	0.61	0.5
MP-H2	0.5	0.48
MP-H3	0.65	0.58
MP-M1	0.7	0.6
MP-M2	0.63	0.6
MP-M3	0.8	0.83
MP-L1	0.7	0.8
MP-L2	0.85	0.8
MP-L3	0.7	0.9
Gmean	1.7	0.55

We have 23% and 42% improvement in read and write access latency, respectively.

Table 1 proves that VR-PCM of 2-bit would be at the baseline by 49% while the average being 23% for read access latency. Moreover, for write access latency the figure shows 72% improvement where the average is 42%. What is amazing is that improvement in performance provided through accessing MFV within VR-PCM is so large that it would compensate the loss in performance while accessing data that is less frequent.

Notably in memory intensive applications, the reduction in memory access time is translated into improvement in system performance. CPI improvement for a multi-core architecture with VR-PCM support is shown in Table 2 while compared to the same type of system utilizing a baseline memory. It also makes it clear about the 26% improvement in system performance (CPI) of 2-bit MLC VR-PCM, which proves write/read access latency reduction.

Table 3 shows the memory read/write energy for VR-PCM and baseline MLC PCM memory. For the energy evaluation, we consider both static and dynamic energy of the cell arrays, peripheral circuitry (shown in Fig. 3), and frequent value processing logic. The normalized result confirms that the VR-PCM outperforms baseline and reduces read and write energy by up to 38% (26% on average) and 42% (18% on average), respectively.

For evaluating the lifetime in our observation, we can assume that the number of writes which are reliable in 2-bit MLC PCM cell can be limited to 10^6 cycles [12] that clearly follows the presented approach in [13]. To simplify the lifetime analysis, we use a perfect wear-leveling like Start-Gap [14]. For different cell storage density levels, our results in Table 4 prove that for 2-bit MLCs, memory lifetime improvement is about 2.61 × (on average 1.71 ×). Likewise energy and access latency, lifetime improvement is expected when VR-PCM is used in workloads exhibiting more value locality.

6.2 Extended vs reconfigurable VR-PCM for high-density MLCs

As previously stated, for high-density devices, more than 2-bits per cell, VR-PCM model needs modification to work better. Extended and Reconfigurable VR-PCM were proposed to ensure the limitations of VR-PCM are overcome when the top values' coverage in frequent value stack wouldn't be too large at 4 bit or 3 bit data granularities.

Figs. 4 and 5 compare average memory access latency and system performance (CPI) of Extended VR-PCM and Reconfigurable VR-PCM for 3-bit and 4-bit MLCs. Compared to the conventional 3-bit MLC PCM main memory; there is 16% improvement in main memory access

Table 2 Performance improvement of a system with 2-bit VR-PCM main memory normalized to conventional MLC design.

Workloads	Normalized cycle per instruction (CPI)
Blackschole	0.85
Bodytrack	0.83
Canneal	0.85
Deadup	0.85
Facesim	0.78
Ferret	0.81
Fluidanimate	0.82
Freqmine	0.89
Raytrace	0.7
Streamcluster	0.85
Swaptions	0.83
VIPs	0.9
X-264	0.95
MP-H1	0.8
MP-H2	0.78
MP-H3	0.78
MP-M1	0.85
MP-M2	0.82
MP-M3	0.85
MP-L1	0.98
MP-L2	0.9
MP-L3	0.96
Gmean	0.84

This chart shows 13% improvement in system's CPI on average.

latency and an average of 12% improvement in CPI for Extended and Reconfigurable VR-PCM. When a specific workload has less value locality, Reconfigurable VR-PCM beats the Extended VR-PCM with an improvement of 3–4% in measured CPI.

Table 3 Normalized read, write, and total energy consumption of 2-bit VR-PCM main memory.

Workloads	Read access	Write access	Total access
Blackschole	0.82	0.9	0.9
Bodytrack	0.8	0.83	0.83
Canneal	0.61	0.82	0.82
Deadup	0.73	0.85	0.85
Facesim	0.77	0.62	0.62
Ferret	0.77	0.62	0.62
Fluidanimate	0.7	0.78	0.78
Freqmine	0.8	0.9	0.89
Raytrace	0.78	0.5	0.6
Streamcluster	0.79	0.82	0.82
Swaptions	0.7	0.85	0.85
VIPs	0.8	0.82	0.82
X-264	0.78	0.9	0.9
MP-H1	0.63	0.6	0.6
MP-H2	0.71	0.73	0.73
MP-H3	0.62	0.61	0.61
MP-M1	0.81	0.8	0.8
MP-M2	0.81	0.86	0.86
MP-M3	0.81	0.82	0.82
MP-L1	0.78	0.82	0.8
MP-L2	0.85	0.9	0.9
MP-L3	0.7	0.82	0.8
Gmean	1.75	0.77	0.77

The results reveal that we can expect considerable energy reduction in workloads with high value locality.

The same trend as 3-bit devices is followed in devices of 4-bit for memory access time and CPI. Thus, our observation reveals that we have 21% improvement in memory access latency resulting in 10% average improvement in CPI.

Table 4 Orders of lifetime improvement when using VR-PCM in a 2-bit MLC main memory.

Workloads	**Orders of lifetime improvement**
Blackschole	1.7
Bodytrack	1.5
Canneal	1.8
Deadup	1.5
Facesim	2.5
Ferret	2.7
Fluidanimate	2.5
Freqmine	1.2
Raytrace	5.1
Streamcluster	1.6
Swaptions	1.7
VIPs	1.7
X-264	1.1
MP-H1	2
MP-H2	2.3
MP-H3	2.1
MP-M1	1.2
MP-M2	1.1
MP-M3	1.9
MP-L1	1.1
MP-L2	1.1
MP-L3	1.2
Gmean	1.70

Here, we experienced an average of 1.71 × improvement in lifetime (5.1 × for raytrace). One can expect more lifetime improvement when the program has large frequent value locality.

Figs. 6 and 7 show that as similar to 2-bit MLC VR-PCM, the energy consumption improvement and lifetime of VR-PCM are very great. It is the case when the workloads would be high-value-local. But, for applications with low-value-local, Reconfigurable VR-PCM has more chance

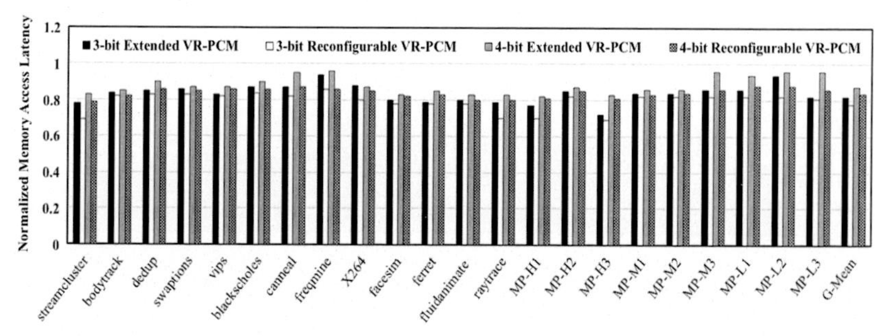

Fig. 4 Normalized memory access latency of 3-bit and 4-bit Extended/Reconfigurable VR-PCM. This experiment confirms considerable reduction in memory access latency especially in Reconfigurable design.

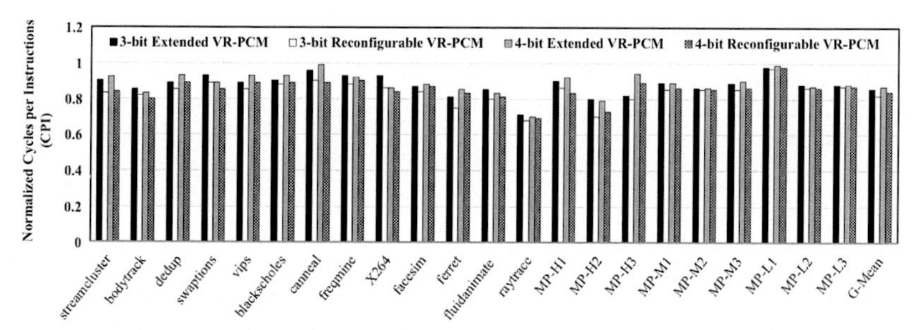

Fig. 5 Performance improvement of a system with 3-bit and 4-bit Extended/Reconfigurable VR-PCM main memory normalized to the conventional MLC main memory design. This chart shows 10% improvement in overall CPI, on average.

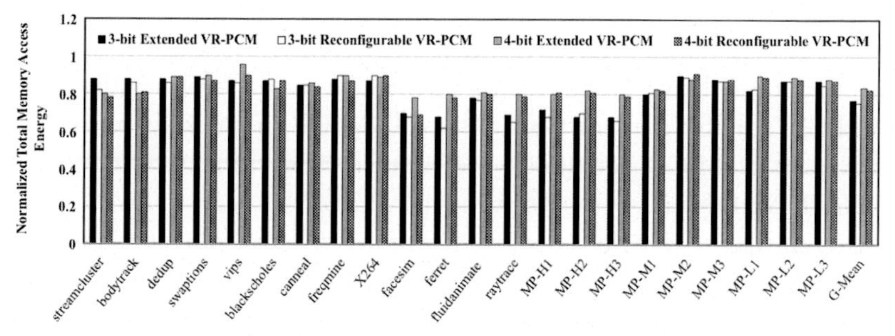

Fig. 6 Total memory access energy of 3-bit and 4-bit Extended/Reconfigurable VR-PCM main memory normalized to the conventional PCM.

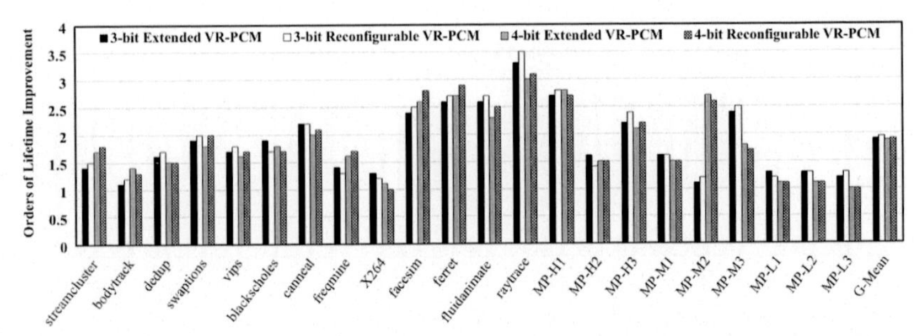

Fig. 7 Orders of lifetime improvement when using Extended/Reconfigurable VR-PCM in a 3-bit and 4-bit MLC main memory. Here, we experienced an improvement of 1.04 × to 3.5 × in lifetime.

to increase the memory performance. Overall, for high-density devices (3-bit and 4-bit MLCs), the Reconfigurable VR-PCM has an improvement between 6% and 35% in main memory energy consumption. Adding to this, the lifetime improvement for the similar memory structure would vary from 1.06 × to 3.5 ×.

7. Process variation and resistance drift

To study the effects of process variation on PCMs [15], process-supply voltage-temperature (PVT) variation effects on cell's geometry and nMOS access transistor have been considered. It is also shown how PVT affects both amplitude and duration of the minimum required electrical current *I*RESET-Min which is used in write process to reset a cell.

The maximum resistance value of the cell (about 1 MΩ) is shown as *I*RESET-Min. It is proven that *I*RESET-Min would be dependent on various variations like: (1) BECD or bottom electrode contract diameter, ThickGST or thickness of GST, ThickHEATER or thickness of heater. As a result of BECD variation, proof from the experiment confirms that *I*RESET-Min deviation can be as much as 32%.

Higher *I*RESET-Min is the product of thinner heater layer and its variation has the potential to affect *I*RESET-Min by as much as 46%. Although *I*RESET-Min isn't sensitive to ThickGST but it happens with about 1% variation for *I*RESET-Min.

Moreover, as the transistor gate length increases because of the increased transistor ON-resistance, the current magnitude can reduce considerably. Thus, it is clear that cell's geometries variation and also nMOS access transistor can impose up to 40% a maximum variation to *I*RESET-Min (Min = 0.180 mA, Max = 0.420 mA, $\mu = 0.300$, $\delta = 0.033$, $\delta/\mu = 11\%$ with the experimental setting in [15]).

The required minimum current for resetting a cell varies across the memory array due to variation in PCM cell's parameters. To successful programming of a PCM cell, write current generator has to be adjusted based on the worst case, i.e., it is adjusted to the maximum of the minimum required currents to program different cells. For proper resetting of all PCM cells (maximum resistance value), write driver functional block would be required to raise the current up to the highest level to enable sufficient current. It would have power overhead dissipated by cells which are affected less by PVT. Each cell could be reset to the maximum resistance value (1 MΩ), up to minimum requirement as it is guaranteed. Later P&V process output would create resistance band which guarantee resistance value requirement as it gets converted through a cell, either in conventional MLC PCM architecture through using uniform resistance partitioning or by employing VR-PCM with non-uniform resistance partitioning.

However, it is assumed that all cells in entire PCM would have the minimum RESET resistance value of 1 MΩ, and the greatly shrunk resistance region which are fully crystalline in VR-PCM would have length of 124, 62, and 31 KΩ *in 2-, 3- and 4 bit* MLC, respectively (the values are, respectively, 220, 115, and 48 KΩ in 2-, 3-, and 4-bit baseline MLC PCMs).

Compared to the wider resistance region in the baseline MLC PCM, narrower resistance region needs more iterations for P&V in the programming process. Wider resistance region in VR-PCM requires less P&V iterations. P&V process would have an average iterates of 13.3 $\times$, 16.51 $\times$, and 19.3 $\times$ for a cell to program to the most shrunk resistance region which is in VR-PCM 2 -, 3-, and 4-bit, respectively (8.7, 12.78, and 18.5 iterations in baseline 2-, 3-, and 4-bit MLC PCMs).

To conclude, RESET current in the worst-case value is similar in both the baseline memory and also VR -PCM for PVT effects to be overcome. However, at amorphous side, wider resistance regions in VR-PCM require less iterations and energy. Conversely, narrower resistance regions at crystalline side in VR-PCM require nonetheless more P&V iterations and consume high energy. By looking at both resistance region assignment and data locality in VR-PCM, it provides less power consumption for both read and write accesses along with having a lower access latency because of faster reads.

7.1 Analysis of drift tolerance

It should be noted that the main feature of VR-PCM is the interstate noise margin reduction in an MLC cell. This narrower noise margin in VR-PCM would consume less energy as it reduces the number of integrations required during P&V. But still this narrow noise margin might be insufficient to avoid the overlapping of states from the unwanted resistance drift. If the final design could tolerate the resistance drift leading to soft errors, VR-PCM would be energy efficient and acceptable. To analyze MLC PCM reliability, we show the bit error rate of the readout data in Fig. 8 for the evaluated

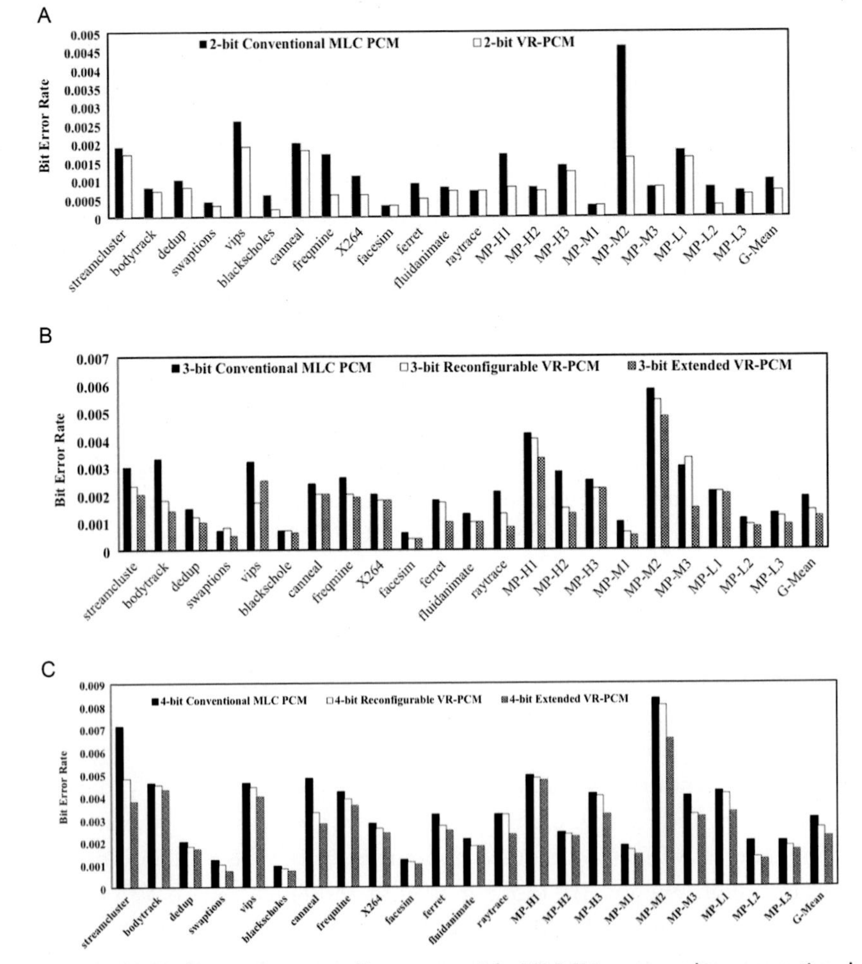

Fig. 8 Rate of soft error (in terms of bit error rate) for VR-PCM compared to conventional MLC PCM design. These charts along with Table 1 confirm that VR-PCM gives same level of soft-error reliability with average of 18% reduction in MLC PCM write energy.

structures in Section 5 for various workloads. Indeed, using the resistance partitioning model used throughout analysis, the conventional baseline memory and VR-PCM MLC designs tolerate about 1.2 s and 1.23 s drift at 300°K, respectively. As it is shown in this figure, the same level of drift reliability with an average of 16.1% reduction in writes' energy is presented by VR-PCM and its alternative designs.

References

[1] Asadinia M, Arjomand M, Sarbazi-Azad H: Variable resistance spectrum assignment in phase change memory systems, *IEEE Trans VLSI Syst* 23(11):2657–2670, 2015.
[2] Zhou P, et al: A durable and energy efficient main memory using phase change memory technology. In *Proc. IEEE Symp. High Performance Computer Architecture (HPCA'09)*; 2009, pp 14–23.
[3] Bienia C, Li K: PARSEC 2.0: a new benchmark suite for chip-multiprocessors. In *Proc. Workshop on Modeling, Benchmarking and Simulation (MoBS)*; 2009.
[4] Joshi M, et al: Mercury: a fast and energy-efficient multi-level cell based phase change memory system. In *HPCA*, 2011.
[5] Awasthi M, et al: Efficient scrub mechanisms for error-prone emerging memories. In *Proc. IEEE Symp. High Performance Computer Architecture (HPCA)*; 2012, pp 15–26.
[6] Bedeschi F, et al: A bipolar-selected phase change memory featuring multi-level cell storage, *IEEE J Solid State Circuits* 44(1):217–227, 2009.
[7] Zhang Y, Yang J, Gupta R: Frequent value locality and value-centric data cache design. In *Proc. Int. Conf. Architectural Support for Programming Languages and Operating Systems (ASPLOS)*; 2000, pp 150–159.
[8] Spradling CD: SPEC CPU2006 benchmark tools, *ACM SIGARCH Comput Arch News* 35(1):130–134, 2007.
[9] Chen YC, et al: Ultra-thin phase-change bridge memory device using GeSb. In *Electron Devices Meeting*; 2006, pp 1–4.
[10] Nirschl T, et al: Write strategies for 2 and 4-bit multi-level phase-change memory. In *Proc. IEEE Int'l Electron Devices Meeting*; 2007, pp 461–464.
[11] Qureshi MK, Franceschini MM, Lastras-Monta LA: Improving read performance of phase change memories via write cancellation and write pausing. In *Proc. IEEE Symp. High Performance Computer Architecture (HPCA'10)*; 2010, pp 1–11.
[12] ITRS: *Process Integration, Devices, and Structures, 2011, ITRS: International Technology for Semiconductor. 2011 edition. Retrieved in April 2012 from, http://www.itrs.net/.*
[13] Lee BC, et al: Architecting phase change memory as a scalable DRAM alternative. In *Proc. Int'l Symp. Computer architecture (ISCA'09)*; 2009, pp 2–13.
[14] Wu Q, et al: Using multi-level phase change memory to build data storage: a time-aware system design perspective, *IEEE Trans Comput* 62:2083–2095, 2012.
[15] Zhang W, Li T: Characterizing and mitigating the impact of process variations on phase change based memory systems. In *Proc. IEEE/ACM Int. Symp. Microarchitecture*, pp 2–13.

Further reading

[16] Zhang W, Li T: Helmet: a resistance drift resilient architecture for multi-level cell phase change memory system. In *Proc. IEEE/IFIP Int. Conf. Dependable Systems and Networks (DSN)*; 2011, pp 197–208.

About the authors

Marjan Asadinia received her Ph.D. degree in computer engineering from Sharif University of Technology, Tehran, Iran, in 2016. During the Ph.D., she worked on Phase Change Memories at HPCAN laboratory. She then acquired a research associate position in the Department of Computer Science at Oregon State University working on advanced technologies for non-volatile memories. She is currently a postdoctoral fellow at the University of Arkansas, Computer Science and Engineering Department. Her research interests include NoC, high-performance computer architecture, and new memory technologies.

Hamid Sarbazi-Azad received his Ph.D. in computing science from the University of Glasgow, Glasgow, United Kingdom, in 2002. He is currently professor of computer science and engineering at Sharif University of Technology and heads the School of Computer Science, Institute for Research in Fundamental Sciences (IPM), Tehran, Iran. His research interests include high-performance computer/memory architectures, NoCs and SoCs, parallel and distributed systems, performance modeling/evaluation, and storage systems, on which he has published about 400 refereed conference and journal papers. He received Khwarizmi International Award in 2006, TWAS Young Scientist Award in engineering sciences in 2007, and Sharif University Distinguished Researcher awards in years 2004, 2007, 2008, 2010, and 2013. He is now an associate editor of ACM Computing Surveys, Elsevier Computers and Electrical Engineering, and CSI Journal on Computer Science and Engineering.

Printed in the United States
By Bookmasters